GriffithREVIEW27
FOOD CHAIN

Edited by Julianne Schultz

GriffithREVIEW 27

Praise for Griffith REVIEW

'Admirable insights…witty and cuts through the complexities.'
Australian Book Review

'Rigorously intelligent and far from self-satisfied.'
Peter Pierce

'Rare, sometimes shocking candour…I recommend a long and leisurely
reading. This collection shows Australia and Australians from every
point-of-view but the obvious, revealing a diversity of worlds.'
M/C Reviews

'One of the few places in Australian media where
arguments can be developed and complexity teased out.'
Margaret Simons

'*Griffith REVIEW* has become important reading, combining
thoughtful analysis and readable writing in an effort to counteract
the all-enveloping spin that blurs reality.' *Weekend Australian*

'*Griffith REVIEW* is a wonderful journal. It's pretty
much setting the agenda in Australia and fighting way
above its weight…You're mad if you don't subscribe.'
Phillip Adams

'The indispensable read for literate Australians.'
Geraldine Brooks

'*Griffith REVIEW* takes academic journalism out of
the ivory tower onto the street and into the countryside…
a refreshing and invigorating move.' *Courier-Mail*

'*Griffith REVIEW* represents "the long game" in journalism,
providing quality analysis in an age of diminishing
journalistic integrity.' *Walkley Magazine*

SIR SAMUEL GRIFFITH was one of Australia's great early achievers. Twice the premier of Queensland, that state's chief justice and the author of its criminal code, he was best known for his pivotal role in drafting agreements that led to Federation, and as the new nation's first chief justice. He was also an important reformer and legislator, a practical and cautious man of words.

Griffith died in 1920 and is now best remembered in his namesakes: an electorate, a society, a suburb and a university. Ninety-six years after he first proposed establishing a university in Brisbane, Griffith University, the city's second, was created. His commitment to public debate and ideas, his delight in words and art, and his attachment to active citizenship is recognised by the publication that bears his name.

Like Sir Samuel Griffith, Griffith REVIEW is iconoclastic and non-partisan, with a sceptical eye and a pragmatically reforming heart and a commitment to public discussion. Personal, political and unpredictable, it is Australia's best conversation.

Griffith
UNIVERSITY

www.griffithreview.com

Read more FOOD CHAIN articles online

Cover photo by W.H. Chong, who was also responsible for the recent redesign of Griffith REVIEW. He is the author of the blog *Culture Mulcher*: blogs.crikey.com.au/culture-mulcher

Griffith REVIEW gratefully acknowledges the support and generosity of founding patron Margaret Mittelheuser.

GriffithREVIEW27 AUTUMN 2010
GriffithREVIEW is published four times a year by Griffith University
in conjunction with Text Publishing. ISSN 1448-2924

Publisher	Marilyn McMeniman AM
Editor	Julianne Schultz AM
Deputy Editor	Erica Sontheimer
Picture Editor & Production Manager	Paul Thwaites
Associate Editor	David Winter, Text Publishing
Publication & Cover Design	W.H. Chong & Susan Miller, Text Publishing
Text Publishing	Michael Heyward, Emily Booth, Sarina Gale, Kirsty Wilson, Ed Austin
Proofreader	Andrea Lewis
Administration	Andrea Huynh
Typesetting	Midland Typesetters
Printing	Ligare Book Printers
Distribution	Penguin Australia

Contributions by academics can, on request, be refereed by our Editorial Board.
Details: www.griffithreview.com

GRIFFITH REVIEW
South Bank Campus, Griffith University
PO Box 3370, South Brisbane QLD 4101 Australia
Ph +617 3735 3071 Fax +617 3735 3272
griffithreview@griffith.edu.au www.griffithreview.com

TEXT PUBLISHING
Swann House, 22 William St, Melbourne VIC 3000 Australia
Ph +613 8610 4500 Fax +613 9629 8621
books@textpublishing.com.au www.textpublishing.com.au

SUBSCRIPTIONS
Within Australia: 1 year (4 editions) $99.80 RRP, inc. P&H and GST
Outside Australia: 1 year (4 editions) A$149.80 RRP, inc. P&H
Institutional and bulk rates available on application.

FEEDBACK AND COMMENT www.griffithreview.com

CAL Cultural Fund — Griffith REVIEW recieves project sponsorship from Copyright Agency Limited.

Australian Government

Australia Council for the Arts

This project has been assisted by the Australian Government through the Australia Council, its principal arts funding and advisory body.

We are what we eat

Food in the time of climate change

Julianne Schultz

WHEN I was growing up, in the 1960s, the food we ate and its supply was tangible – literally outside the dining room window.

We had cows for milk and cream; sheep that grew from suckling lambs to Sunday lunch, and almost every other meal of the week; chickens we nursed in front of the fire, until they became chooks whose eggs we ate and whose feathers we plucked, when their recently headless bodies stopped the mad dervish dance down by the woodpile; vegetables that still had clods of dirt on them when they reached the kitchen; and fruit that fell from the trees, but for most of the year came from Fowlers Vacola jars deep in the pantry, an out-of-season topping for cereal or the base of the nightly dessert.

There was nothing sentimental about this. Our animals were not pets – they were creatures that fed us, and that could be trucked to the saleyard for a few dollars to pay pressing bills. It was smelly, dirty, unrelenting hard work, even on the fertile plains of Victoria's Western District.

Most of the time we ate what my father produced, and my mother cooked. We did not think that we were fortunate; it was the way life was. Food came from the ground; it was seasonal, predictable and, apart from the occasional pavlova or brandy snap, pretty boring. Even our city cousins had chooks in their yards – those in the southern states had trees laden with mandarins, apples, almonds, plums and apricots, and the northerners had backyard trees that produced mysterious soft-skinned mangoes, pawpaws and bananas.

Occasionally we glimpsed another world. Family friends, who owned a big farm nearby, would load us into their two-toned, finned Chevrolet and take us to town, seven miles up the road. It was a special treat and signalled that the cheque for their fine wool had arrived.

There in Penshurst's main street, in the dark and dusty milk bar, they would buy us Chiko rolls, Violet Crumble bars, Cheezels and Tarax soft drinks and laugh, in a kindly way, as we devoured this exotic food with an alertness to texture, flavour and packaging worthy of a Michelin Guide assessor. The crispness of the batter and the ooze of filling in the Chiko roll was unlike anything we ate at home; the tongue fizz of the Violet Crumble bore no relation to the honeycomb we made with golden syrup and bicarb soda; the sharp bubbles of the soft drink were shockingly different to the fresh orange juice we drank, without second thoughts, every day. Cheezels were clearly morsels from another world…

I recall furtive conversations later at home in the manse with my sister. 'This must be the food rich people eat; if only we lived in the city we could eat this stuff all the time.'

We were wrong, of course.

Rich city people are much more likely to want to consume the food we grew up with – local, seasonal and organic. Poor people – including poor rural people, who make up a disproportionate number of the total – are much more likely to eat the cheap, mass-produced and packaged sustenance sold in convenience stores.

It is nonetheless easy to understand our misapprehension.

Entreaties to eat what was on the plate, and think about the starving children in India, rang in our ears. Every year we enthusiastically engaged in Freedom from Hunger drives that filled the church-hall kitchen with countless tins of powdered milk that disappeared to mysterious destinations. ('Do poor people in India need Nestlé's powdered milk because cows are sacred for them?' we wondered.)

The disconnection between food production and consumption, between the food available to the rich and the rest, is now a matter of global anxiety. It is set to become more pronounced as the world's population soars to nine billion and global warming disrupts traditional weather patterns. The Food and Agriculture Organization estimates that there are at least nine hundred million people without enough to eat every day. Even in developed countries, despite an epidemic of obesity, a shockingly large number of people go hungry – forty-nine million people in America alone in 2009.

The enormousness of the problem makes it hard not to be pessimistic. A more enlightened approach to feeding the world's hungry – by giving them the tools for sustainable production rather than having them wait for shipments of instant noodles or powdered milk – is producing impressive results, but the challenge looms like a threat.

Food production and food security will be a bellwether of climate change. There is a lot of talk about rising sea levels, energy costs and extreme weather, but the price and supply of food will bring the reality home. As more than half the world's population now lives in cities, urban food production will overtake the rural-peasant allotment of old.

After decades of detaching food production from the cities, all the major centres now have well-advanced plans to create city gardens. This is a throwback to the time when what are now inner-city botanical gardens were the food source of the colony, and regulations are being loosened to allow backyard chooks and encourage people to plant nature strips with fruit-bearing trees, herbs and vegetables.

Even in Australia, a country with abundant supplies of quality food, anxiety is growing. Talk about water allocations and licences masks the truth of a diminishing harvest from much of the Murray-Darling. Senator Bill Heffernan's advocacy of the deeply embedded dream of a northern food bowl is less the reassertion of a national fantasy than a farmer's intimate understanding of the connection between production and consumption. The ambitious plan to make Tasmania the new national food bowl is the product of original thinking not constrained by established verities.

WHEN MY FAMILY moved to the city, in 1971, we were unwittingly a part of the final phase of the great economic and social transformations of Australia from an agricultural economy to something else. At the time, baby boomers thought that they were escaping to a more interesting life in the city. Their parents could see that the days on the land were numbered – that an era was coming to an end, and the future was uncertain.

The relocation to the city was a consequence of the upheaval in agricultural production – farms were aggregated and production professionalised;

the supply chains from paddock to supermarket became more clearly defined. As a result, the nation's food could be produced much more efficiently and be shipped vast distances and still be fresh and affordable. There was no longer a need for so many people to collect their own eggs, chop the heads off their chooks or milk their own cows.

A country which had made its wealth by shipping its agricultural products to the world, which rewarded its returning soldiers with (suboptimal) farmland and which grounded its sense of identity in the bush, was about to become unequivocally urban. The rural economy was no longer the backbone of the nation and, as farms aggregated, towns faded away and the gap between the country and the city widened.

The pressures of climate change and population growth are making people ever more aware that something has been lost in industrial food production. Now, agriculture accounts for less than 5 per cent of GDP – about the same as the creative industries.

If we are what we eat, Australia is profoundly different to the country of my '60s childhood. These days we eat food that is grown here, much of it processed by large transnational companies with headquarters in Japan, Europe and America. We buy it from supermarkets and grizzle about the prices, and wonder if this is sustainable and flock to farmers' markets. We book out the best restaurants and churn through a global village of ethnic cafés. Schoolchildren follow Michelle Obama's and Stephanie Alexander's example and plant organic gardens. Cookbooks sell in extraordinary numbers, and cooking programs win the television ratings.

Now we are on the path to another major transformation – one that reintegrates the production and consumption of fresh local food into much longer food chains. The major chains have started organic food shops, a glamorous reworking of the local greengrocer.

When I see my urbane father navigate his designer trolley through the stalls at the Victoria Market, it is clear that his journey to seek out fresh produce is every bit as visceral as it is for the émigrés manning the stalls – the Greek fishmongers, Italian deli owners, Vietnamese market gardeners.

For my mother, who took to the role of part-time-farmer's wife reluctantly, the combination of fresh food from the market and the supermarket is infinitely more appealing than plucking chooks, or devising yet another meal drawn from a freezer overflowing with every imaginable part of the lamb she had watched us feed with a bottle.

My father's contact with food-producing soil may now be second-hand, and confined to growing a few tomatoes and herbs, but the connection with fresh produce is as essential as it was decades ago – even in the middle of a big city.

23 November 2009

ESSAY

Sustaining a nation

A river journey from basin to bowl

Margaret Simons

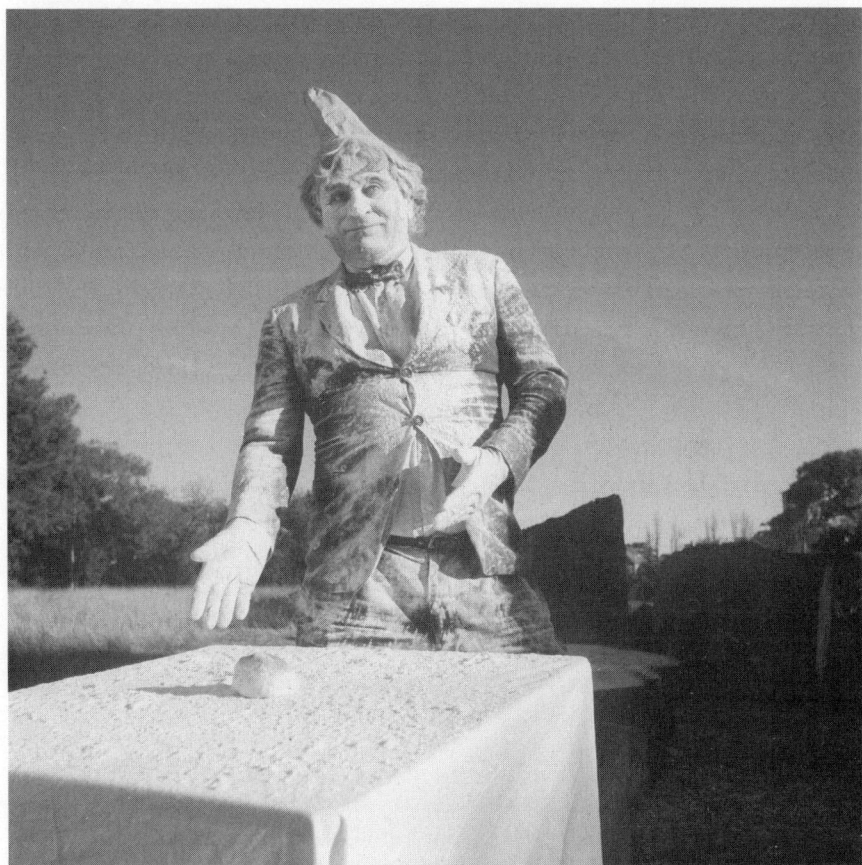

WE visited our relations in spring last year. They raise their living from the central western plains of New South Wales, near Forbes. This is not the kind of country where city folk buy hobby farms, or aspire to holiday homes. A day's drive from Sydney, it is not easy or pretty. It is working-farm country, straightforward and pragmatic, although the flat pastures and waving wheat have their own beauty in good times. These are not good times.

Lambs were fetching four dollars a kilogram at the Forbes saleyards in the week of our visit. That is how they are sold – not on the weight of the bleating animals, but on a calculation of what their carcasses will yield in meat once the hide is removed, the internal organs scooped out, the blood drained away and the head, feet and tail disposed of. Only we city slickers, leaning over the railings of the yard, smelling the manure, watching the animals roll their eyes and push in fear as the auctioneers shout, see a chasm between the cold calculation of 'dressed weight' and the reality of living creatures.

The animals are not only animals. They are units of production. Their entire lives – the costs of rearing, feeding and transporting – are calculated in dollars, with the end point being dressed weight, on which the farmer is paid. According to Meat & Livestock Australia, the costs of production for an efficient lamb producer are around two dollars a kilogram, dressed weight. An inefficient producer might have costs as high as $3.34, leaving little room for profit and what the corporation calls the 'lifestyle aspirations that your farm must support'.

After the dressed weight is calculated and the farmer paid, fat is trimmed. Bone is cut away. More blood is lost. Up to half the dressed weight disappears between sale of the living beast and dinner.

The owner of the farm we visited, Graeme McIntosh, has a saying: 'You've got livestock, you've got dead stock.' There is little room for sentiment, yet they take such care. We heard about how Graeme and his partner, Yvonne, spent hours picking barley-grass seeds out of the eyes of their sheep. The animals have two eyelids, and the sharp seeds get caught in between. The pain can drive animals mad, or, in the cool language of farming publications, lead them to 'lose condition'. To pick out the seeds, Yvonne had to hold the

PREVIOUS PAGE: *Murrumbidgee Jones makes a scone.* Photographer: Michael Shirley from the *Sydney Morning Herald* Shoot the Chef! 2009 photographic competition.

animals down, one by one, warm waxy wool against her work shirt, while Graeme took to them with tweezers.

In the yards near the back door of the farmhouse, Yvonne keeps an enclosure of baby lambs who have lost their mothers. She buys powdered milk in the supermarket, mixes it, and feeds it to the lambs morning and evening in old soft drink bottles topped with a baby's bottle teat. My children love to help with the feeding, although the hungry lambs butt so hard that it can be difficult to hang on to the bottle. Once, the children found an orphan lamb close to death and brought it back to the farmhouse. After a few days of Yvonne's care, it was clearly going to live, and the children named it Lucky. On our most recent visit, they asked what had become of it. Lucky had gone, of course. My husband suggested that next time a better name might be Chop, or Rolled Roast.

I have heard tales of how, many years ago, sheep on a farm near here had to be put down and buried in a mass pit in a time of drought, when they could be neither kept nor sold. The pointless mass killing could have broken hearts but after some hours, as each carcass hit the trench, the people doing the shooting broke into song. It was the song from Monty Python's *Life of Brian*: 'Always Look on the Bright Side of Life.'

When we visited the Forbes saleyards last year – when lambs were being sold for four dollars a kilogram, dressed weight – lamb forequarter chops in Woolworths and Coles cost around eleven dollars a kilo. In between was transport, slaughter, more transport, butchering, the packing into plastic trays and all that is involved in maintaining the bright lights and cold cabinets of the modern supermarket. The profits on any one piece of meat are not big. The two big supermarket chains, Coles and Woolworths, make big money, but largely because of the volume of stuff they sell, rather than enormous mark-ups. The industry relies on volume.

IN PREVIOUS YEARS, I have visited the farm in summer, when the region is consumed in the business of the wheat harvest. Combine harvesters work through the night. Farmers survive with little sleep. Wheat dust covers everything. I have sat in a truck on its way to the silos, which stand silent for most of the year but at harvest time are so busy you have to watch yourself or be run over. Selling wheat is not as simple as it once was, when the grower took the Wheat Board pool price and left it at that. These days, farmers can sell for cash at the silo or take a fixed-grade contract, which can give a premium price but leaves them exposed if the grade or yield of the crop fails to live up to the contract.

There are other buyers at the silos as well: biscuit manufacturers and exporters offering their own prices for different grades of wheat. The truck drivers use mobile phones to keep in touch, and calculations are made on whether to go to one silo or another, factoring in the cost of fuel and driving time. The trucks pull in under the elevated office. Samples are taken from deep within their loads, and measured on the spot for moisture and protein. The higher the protein level, the higher the price. Wheat that contains too much moisture is rejected and must go back to the farm, perhaps to be stored for feed.

All this activity, all this hard work. All the fertiliser, the fuel and the rain. It seems extraordinary to me that at home I can buy a loaf of bread – baked, packed, wrapped and sliced for my convenience – for just $1.80.

There hasn't been a decent wheat crop in Forbes for some years. The Yarrabandai silo is silent and empty. On its side there is ironic graffiti: 'The Hub of the Universe.' There is not a soul to be seen.

The lives of my farming relations are not about the Epicurean, nor about the personal relationships of farmers' markets, nor about any of the fads and fashions of food that haunt the city. My relatives are preoccupied with the hard and gritty business of the industrialised production of food, which remains the means by which most Australians are fed. They are proud of this.

In the city, we risk making a fetish of food. There is so much of it, and it is so cheap. There have never been so many cookbooks, celebrity chefs, unusual ingredients, or attention paid to gradations of taste in oils and nuts, breads and vinegars. There are parts of the city where, for hundreds of metres, every business is a café or a restaurant or a fast-food outlet. In living memory, we have moved from a situation in which most people were worried about not getting enough food, and in which the breadwinner was preoccupied with just that, to one in which most of us worry about how not to overeat.

Yet, while food has become so fancy and fussed over, the growing of it has become less important to our national economy and our national psyche.

In the 1950s, agriculture – mostly food production – made up almost a third of the nation's gross domestic product. Today it is 3 per cent.

Growing food is about connections, but these days it is also about disconnections. The country and the city seem out of sympathy with each other. We visited the farm during the Sydney dust storm of September 2009. Graeme and Yvonne were scathing about the media coverage. In the farmhouse, dust storms are a weekly occurrence. Every nice thing in the house is in sealed plastic bags. The lounge-room chairs are permanently dressed in dust covers and still the fine red grit gets in everything. Reading about what happened in Sydney, Yvonne almost cracked a grin. 'Those poor petals,' she said.

The farm is connected to the cities by roads, but more powerfully by money, and economics, and what is metaphorically described as a supply chain. Every link on the chain is a location of tension, of the playing out of competing interests.

THE AUSTRALIAN COMPETITION & Consumer Commission (ACCC) reported on the competitiveness of the grocery industry in 2008. It examined whether the rising price of food was reflected in the prices paid to farmers, whether the supermarkets were taking unfair advantage. The resulting report was almost five hundred pages long. The supply chains for food are almost impossibly complex and various. 'There is no single story that can be told about the grocery supply chain in Australia,' the report said.

The ACCC undertook a series of case studies on milk, chicken, apples, bread, eggs, biscuits and beef. Each product had a different chain, a different coalition of growers, processers and wholesalers. Some products, such as beef, were aimed mostly at the export market. Others were almost entirely sold through domestic supermarkets, meaning the buyer had significant market power. Suppliers could hardly walk away from negotiations.

The ACCC concluded that the big two supermarket chains, Coles and Woolworths, accounted for about half of all the fresh produce sold in Australia, and nearly three-quarters of packaged goods, yet the industry was still 'workably competitive'. There was keen price competition on the three hundred or so products the supermarkets knew customers used to assess value – including lamb chops and bread. Sometimes that meant these products were sold below cost to bring customers in. Sometimes the sheer volume of their purchases meant the big two were able to squeeze farmers on price. But the ACCC found no consistent evidence that farmers were being taken advantage of. Generally, higher prices in the supermarket were reflected in higher prices at the farm gate, it said. Despite the Horticulture Australia Council saying that 85 per cent of its members felt that growers were unwilling to raise concerns with major retailers for fear of retribution, the ACCC found that the numerous anecdotes and allegations about standover tactics and threats by the big supermarket chains were not reflected in hard, actionable evidence.

And any attempt to alter the relationships, to free up the links in the supply chain, hit obstacles. The ACCC recommended changes to regulations governing the horticulture industry, to make it mandatory for merchants to tell growers what price they would get before the produce was delivered. At present, the price given to the farmer depends on what the merchant can get in a fluctuating market — sometimes high, sometimes low, sometimes nothing at all, with rejected produce being returned. There are no guarantees until the money is in the bank. The ACCC's recommended changes would have transferred the risk of the market from the grower to the merchant, but even this wasn't simple.

The federal government sent the ACCC's recommendations to an industry committee for assessment, but the committee members could not agree. Farmers worried that the changes would mean merchants would offer only very low prices so they would be more able to manage the risk of the market. Premium prices for growers would become a thing of the past. On 1 November 2009, the industry committee's report was put aside for yet more consideration. There is no sign of speedy resolution. Meanwhile, the Minister for Agriculture, Tony Burke, acknowledged that at every link on the supply chain connecting growers to eaters, there are 'considerable tensions'.

Sitting in his dust-proofed lounge room, Graeme McIntosh commented that once, everyone believed that growing food was unquestionably a good thing to do. He wonders whether people in the city still think that. These days, the farm depresses him. Mostly it is the drought. He has grown a beard in the last few years. It is white and trails down to his chest; he has told his family he will cut it off when the drought breaks.

A year ago, last summer, Yvonne took us for a walk over the paddocks and showed us how the soil had opened up in the wheat paddocks – not mere cracks, but great fissures and funny potholes – because all the moisture had disappeared from the soil. 'They never see this in the city,' she said. 'They don't understand.' The wheat crop had failed yet again for lack of rain, and the sheep had been sent in to graze on the stunted remains. In the farmhouse, the floors were like waves because of the shifting clay underneath.

None of the young people in this family plans to take on the farm when Graeme and Yvonne retire, nor are they encouraged to. The pay is too low, and the work too hard.

Our recent visit was a time of hard decisions. Another wheat crop, half grown, was withering. Graeme had to decide whether to hang on in the hope of a little rain, or slash it while it was still good for making hay. He elected to cut his losses, and was out until late every night raking and slashing and raking some more. The drought goes on and on, and his beard gets longer.

Surely, Graeme and Yvonne say, growing food is as important as ever? Surely the drop in the proportion of GDP cannot entirely reflect the value of that often ignored fact, that Australia is one of the few countries in the world with the capacity to feed its own population from its own land, and feed them very well. How much does this matter to us? And what, in these hard times, are we going to have to give up?

GRAEME AND YVONNE'S farm lies in the middle of the Murray-Darling Basin, the food bowl of the nation, and the exemplar of the connections and disconnections in food production. Water connects people. Paul Keating called the Murray-Darling a 'real and symbolic artery'. The metaphor is a cliché when applied to most rivers, yet it is right for the Murray. This river system is what makes a large part of Australian life possible. It is both lifeline and drain, with capillaries running to and from it, carrying water to crops and cities, and discharging waste so it can be carried from the interior to the sea.

The basin is home to more than a third of the nation's farms. It covers about a seventh of Australia, and is responsible for more than a third of its agricultural production. The country's three largest rivers – the Darling, Murray and Murrumbidgee, which water three-quarters of Australia's irrigated land – run through it. The Murray-Darling Basin grows nearly all our rice, most of which is exported. Nearly all of Australia's oranges are grown on Murray-Darling irrigation. In the basin, we raise and grow more than half the nation's pigs and apples and almost half its wheat, as well as grapes, vegetables, nuts, cotton, canola and pasture for dairy cattle, beef cattle and sheep.

In the popular imagination, rivers are places of meaning and memories. The Murray is one of the geographical features that connects the Dreaming stories of the first human occupants. Most of the tribes along the river shared the story of the great Murray cod – Ponde, in the language of the lower Murray – that carved the river's course before being speared by the Dreaming hero Ngurunderi in the lower lakes. It was another river myth about the lower lakes – the so-called secret women's business of the island at the mouth of the Murray – that transfixed the nation in the early 1990s, and set the early Aboriginal-affairs agenda for the incoming Howard government.

The European myths of the river are also powerful – archetypal, even. Europeans, beholding the inland, dreamed of gardens in the wilderness and fruit growing in the desert. Realising these dreams was the job of engineers. In the postwar years, engineers were heroes. The historian Ernestine Hill published in 1958 a history of irrigation, *Water into Gold*. 'Here is the beginning of a great story,' he wrote, 'the transfiguration of a continent by irrigation science…the radiant twin cities of Mildura and Renmark, the Garden Colony in that lucky horseshoe of Murray River that unites two Australian States, will always be our first national shrine to irrigation science.'

Hill described the Chaffey brothers, Canadians who joined pioneering irrigation schemes in California and later developed Mildura and Renmark, as 'apostles of irrigation'. The water they pumped on to the land was a benediction, bringing civilisation to the dusty and hostile Mallee wastes.

And yet something has gone wrong with the story-making and the engineering marvel. We hear it all the time now, and the engineering terms are unintentionally ridiculous. There have been 'dry inflows', meaning no inflows. The river has been 'over-allocated', meaning more water is given away than the land can provide.

TODAY, THE MURRAY-DARLING system is not only a collection of rivers. Rather, it is a feat of plumbing. The Murray-Darling Basin Authority has schematic maps of the system that show it as a canal, with weirs and storages and taps that enable the precise management of water. The dreaming of the river has become separated from the way the water is managed. Engineers manipulate the flow by tapping keys on a computer many kilometres away from the river red gums, the birds and the big brown smell of the water.

It was in the early 1990s that the engineers fell from grace, when a toxic blue-green algae outbreak covered a thousand kilometres of the river, making the water poisonous to man and beast. The algae forced us to realise that this sluggish, harnessed river was not healthy. People began to talk about it dying.

The dream of turning water into gold was questioned. People began to backpedal, and to research. There had been warning signs before, but 1991 marked the big shift in attitude. The late Peter Cullen, the water scientist and founder of the Wentworth Group, saw in the blue-green algae outbreak a change in our understanding of the nation. 'Every now and again, politically, we get a focusing event when the ideas of different people all coalesce and you have a chance to change in this country,' he said.

It quickly became apparent that there simply wasn't enough water. In 1992, the Murray-Darling Basin Commission was set up to manage the basin, although most water matters remained with the states. In 1995, a water cap was put on allocations, and recently the achievement of the Rudd Government was an intergovernmental agreement setting up the Murray-Darling Basin Authority and strengthening the role of the ACCC to oversee the trade in water.

The main responsibility of the new authority is to devise the first comprehensive plan, encompassing both groundwater and river water, to cover the entire basin. It is a massive research undertaking, which will in time affect every windmill creaking on every farm, every drip-irrigation system and every dam. A draft plan will be released this year and the final version will be completed in 2011. Once the plan is adopted, it will be binding across the nation. The states, having transferred their water powers to the Commonwealth, will not be able to veto the caps and management arrangements imposed by the Murray-Darling Basin Authority. For the first time, the basin will be managed as a whole.

At least, that's the idea. There is a lot of room for slippage between conception and reality. There will be big political battles: the states will not let their water go easily. The water of the Murray-Darling Basin may connect us, but in the world's driest continent, it also profoundly divides us.

The management of the Murray-Darling has always been central to the nation. Nearly all of the Murray's flow is generated by the rivers of Queensland, Victoria and New South Wales, yet South Australia depends on the water, too. When the Australian Constitution was being devised, South Australia insisted on a section giving the Commonwealth power to pass laws for navigation and shipping – the main use of the lower river at the time. New South Wales and Victoria responded by inserting a clause stating, 'The Commonwealth shall not…abridge the right of a State or the residents therein to the reasonable use of water for conservation and irrigation.'

Ever since, the debates have turned on that word 'reasonable'. Yet only in 2010 – when the water plan is complete, more than a century after the constitution was drafted – will we have clear data on what that word might mean, and what amounts to sustainable use.

The Murray-Darling Basin is characterised by droughts and floods, but the length and severity of the current drought is testing everything we thought we knew about this land, and highlighting, if it needed to be highlighted, that the most important part of environmental policy is not plastic bags, shower nozzles, recycled cans or bricks in the toilet cistern. It is food. No human behaviour has greater environmental effects than diet, and nothing connects these effects like water.

Since 1917, the sharing of water has been governed by an apparently simple agreement, in which the inflows above Albury are shared equally by New South Wales and Victoria, subject to an obligation to supply South Australia with a fixed entitlement. In times of low inflow, the fixed obligation to South Australia is replaced by a fixed share. Now, under the pressure of 'dry inflows', those agreements are unravelling. Although the states have handed power to the Commonwealth, Victoria in particular is capping the amount of water it is prepared to trade downstream, and South Australia is considering a High Court challenge to force the three upstream states to allow more water to flow. Already there have been rowdy demonstrations in the irrigated districts of South Australia.

Farmers in Victoria and southern New South Wales are disgusted that a pipeline will soon carry 'their' water to Melbourne. 'Why should we suffer because Melbourne is not sustainable?' one farmer said to me. In the Riverina, they hate the idea that 'their' water may be used to keep the lower lakes in South Australia from becoming toxic. In the Murrumbidgee Irrigation District, the farmers resent that some of their water entitlement has been traded downstream. In South Australia, they are aghast that in New South Wales water still runs in open irrigation channels, when we know that more than three-quarters of it can be lost to seepage and sun. Throughout the basin, there is a sense of injustice, envy and fear. Drive through the Murray-Darling Basin today, and you get the feeling that water wars – the stuff of science fiction – may not be very far away.

WHEN WE LEFT the farm at Forbes last spring, we decided to follow the course of the Lachlan River, which on the map is shown, in a deceptively clean blue line, as a tributary of the Murrumbidgee. In fact, its water rarely makes it to central New South Wales. In a normal year, whatever that means these days, the Lachlan ends in marshes. In 1817, the explorer John Oxley followed the same course we drove, hoping to find a river that flowed the width of the continent, that would open up the inland to navigation and development. He made the journey in a wet season, and found only, as the historian Michael Cathcart recorded in his recent book, *The Water Dreamers* (Text, 2009), 'sickly marshes and unhealthy plains' where the turbid waters spread 'silence, death and desolation'.

Oxley was wrong. When this country was drained, it became prime farming land. Yet today, without water, there is once again silence, death and desolation.

We hugged the course of the river, following dirt roads. The country was dun-coloured, the only farm activity the slashing of failed wheat crops for hay. Sometimes a crop had not even made it that far, and the sheep and cows had been allowed in to graze. What did that do, I wondered, to the calculation of the cost of production?

We drove to the riverbank, and scrambled down past trees to reach the green-brown flow. The roots of the trees were splayed out above us like arthritic hands, groping for the water that would normally cover them. In the week we were there, the irrigators in the lower reaches of the Lachlan were cut off. There were to be no more irrigation diversions that season.

We wearied of the drab depression of the drive, and decided to strike south. After only an hour, we were in a different world. The Murrumbidgee Irrigation District was first a deep-green splash on the horizon, then all around us. It was almost a caricature of plenty. On either side of the road, grape vines carried ripening fruit. The orange trees were laden, and more fruit lay on the ground.

We drove in to Griffith and had a picnic on grass as green as paint, alongside the main irrigation channel, which runs, wide as a city street, through the heart of town. We even saw a council worker hosing down a footpath. There were sprinklers watering lawns in the midday sun. We were shocked, and yet our spirits lifted. The dream of the settlers and the engineers, the garden in the desert, is still alive in Griffith, and it was hard not to relax into it, to enjoy the cornucopia.

In the centre of town, more prominent even than the war memorial, was a monument to an engineering invention, the Dethridge wheel. Named after its inventor, a commissioner of Victoria's water supply, the wheel is a drum on an axle with V-shaped veins fixed to the outside. John Dethridge refused to patent his invention, instead offering it for the benefit of humankind. It was taken up all over the world. Placed in an irrigation channel, the Dethridge wheel is used to measure the water provided to each farm.

In irrigation districts, you can hear plenty of stories about how farmers used to beat the wheels and steal water. A favourite trick was to jam a wheel with a frozen fish. By the time the fish thawed and the wheel turned again, thousands of unmetered litres would have flowed to the farm. In Far North Queensland, the story goes that they used dead turtles instead of fish. The inspectors knew the tricks, but how to prove that the fish or turtle had not died a natural death and been stuck by accident? The irrigators tell these stories with the same spirit that city folk talk of speeding fines, or computer hackers talk of free downloads. Water theft has only recently been regarded as a serious crime.

Griffith and its surrounds is, more than any other place in Australia, the product of the engineers' dreams. The town even takes its name from a New South Wales minister for public works. It owes its existence to the biggest water dreaming of all, the Snowy Mountains Scheme. More water than Sydney uses in a day flows each day from the scheme into the Murrumbidgee and Tumut rivers. From there, it is diverted into the main irrigation canal and carried 155 kilometres, then through another 2,350 kilometres of channels to farm boundaries.

One of the oldest farms in the district, Catania, runs a tourist sideline showing visitors around the block, then selling them homemade jam and wine, and olive oil smelling of freshly cut grass. The farm makes prunes out of its plums, but these days some of the fruit is exported fresh to Japan, where fresh 'sugar plums' sell for more than forty dollars each. Those are the kind of prices you have to pay, the farm owner tells the tourists, if your country lacks the ability to grow its own fresh food. We were told that the produce of the district had repaid the country many times for the investment of taxpayer dollars that had made the irrigation scheme possible.

There are problems in paradise, though. The first impression of plenty is partly deceptive. Water is allocated through a system of high-security and low-security licences. High-security licence holders – mostly the growers of permanent crops like fruit trees and vines – get a fuller allocation even when the supply is short.

Low-security water is used for rice and wheat. Half an hour's drive from Griffith is Leeton, the centre of the Australian rice industry. Next to the Sunrise factory, there is a small tourist shop where you can watch videos about how Australia's rice crop helps feed the world. Here you can find brown rice, organic rice, Arborio rice and long-grain rice. You can buy twenty-five different flavours of rice cracker. Yet none is grown in the district. They have had to change the labels on the packets from 'Product of Australia' to 'Product of Thailand'. Around the factory, the rice fields are empty.

Like every other water user in the basin, the rice industry is investing money to convince the public that it is an economical consumer of water. Certainly, the pamphlets say, rice must be flooded, but after the harvest another crop – wheat or canola – will be planted in the still damp soil.

You see these kinds of claims a lot in this time of 'dry inflow'. The beef industry disputes the figures provided by conservationists that say steak carries more 'water debt' than any other food. Horticulturalists talk about investing in drip irrigation, and claim that vegetables are water-wise. In truth, the 'water debt' of food is hard to measure, and depends on many variables, but the CSIRO issues heavily qualified figures that estimate that each kilogram of maize takes up to 630 litres of water to grow. Rice takes 1,550 litres, and beef between 50,000 and 100,000 litres.

In raw terms, the biggest water users in the basin are cotton, dairy farming and pasture, and rice. Together they account for almost three-quarters of agricultural water consumption. Fruit and vegetables are much further down the list.

But in a time of water scarcity, the figures that really matter are the dollar returns for each megalitre of water used. These are the numbers that frighten the rice growers of Leeton. These are the figures that have caused each state, and each region, to resist the dictates of the market. Rice returns just $200 for each megalitre used. Livestock, pasture and grains return $300, and cotton $600.

If we were to use water in the most economically rational fashion, we would grow vegetables ($1800 for each megalitre of water used), fruit ($1500) and grapes ($900). Such a decision would wipe out large communities throughout the basin – the cotton growers of Queensland, the rice growers of New South Wales, the dairy farmers of Victoria – with all the dislocation that entails.

If Australia were governed by wise dictatorship, there would be huge forced land-use changes in the Murray-Darling Basin. Some crops would not be grown. Some communities would be relocated. State governments would be forced to comply. Instead, the effort to save the basin is a matter of slow and uncertain negotiation, limited by our awkward federation. It is also a matter of the imperfect instrument of market forces.

Water trading is one of the main ways to mitigate the effects of drought. Over the past decade, enormous effort has gone into unravelling the byzantine historical rules and regulations that govern the allocation of water. It has been directed at uncoupling water rights from land, to allow the water to be traded between properties and even between states. The process has been extraordinarily complex. The means by which water is owned and allocated vary from state to state and region to region. A farmer in Griffith will probably not understand the system in Mildura, and vice versa. Nevertheless, water is traded from one to the other and sold like any other commodity. The idea is that if the market is allowed to rule, water will naturally find its level and its most highly valued use.

Water trading is not a simple answer, nor a perfect one. Water does not flow like money. It can only be traded when there is a physical connection between seller and buyer, and when the buyer is downstream. There are good reasons to have reservations about the unnaturalness of uncoupling water from land. If water is traded out of a region, the remaining irrigators will be left to pay for irrigation infrastructure. This alone could send otherwise economic farms broke.

Additionally, water having a separate tradeable value means that many 'sleeper' irrigation licences that were not exercised are suddenly valuable. Water that previously never left the river is now for sale.

Finally, there is the fear that water barons will buy up the river and take it to new locations, leaving the family farmers high and dry. Is it wise, the farmers ask, for this most precious resource to be in the hands of people motivated only by profit?

The assessments on water trading have so far concluded that it has helped the basin and its people to deal with drought. The effects would have been worse had water remained tethered to land.

Yet it is not easy to let the water go. The CSIRO predicted that the basin would have to deal with up to a third less water by 2030 due to climate change. Regions such as Griffith and Leeton can be expected to resist the logic of the market. In the last annual report of Murrumbidgee Irrigation, its chair of the board, Dick Thompson, urged shareholders not to trade their water. 'It is unfortunate from a regional perspective that we have seen water move from highly desirable and essential production with larger common benefits…to support overproduction and unviable production elsewhere. I would urge all shareholders to seriously consider producing a crop with any water they have, not only in their longer-term interest but also the community's.'

From the perspective of an irrigator, every use downstream appears less pressing, less worthy, than use in their own district. I asked the growers of Griffith why their orchards were still productive when downstream, in the Riverland of South Australia, trees were being ripped up and grapevines left to die. Their response was somewhere between puzzlement and belligerence.

South Australia should be all right, I was told. After all, despite their reservations, they had traded water downstream. If the South Australians couldn't make it work, that was their problem.

Meanwhile, the federal government is in the market, with three billion dollars to spend on water to return it to the river. The big purchases, such as the huge Toorale Station on the Warrego and Darling rivers, make the headlines. With the departure of the water, crops and people must go, too. Some of the land returns to dry-land wheat farming. Other farms go out of production altogether. In this way, policy changes the landscape, and the economies of whole districts, while rerouting the water like a careful gardener to the wetlands downstream, trying to save the soils from becoming acid, and the birds, red gums and bell frogs from dying.

It is acknowledged that until the Murray-Darling Basin plan is completed, the purchases are being made on the basis of inadequate information about where water can most effectively be returned. The federal government has rejected the idea of compulsory acquisition. It will deal only with willing sellers, rather than forcing sales in areas where they could do most good. Yet what else can be done?

Penny Wong, the federal water minister, said recently, 'When dealing with a patient who is critical, you have to act. I could have spent years in the courts arguing over a perfect outcome, and hoping the river was still there when I got out of court. Or I could get on with the job.'

A RECENT BOOK by Jessica Weir, *Murray River Country: An Ecological Dialogue with Traditional Owners* (Aboriginal Studies Press, 2009), records that the Aboriginal communities along the river have a dream that one day there will be a great flood – a flush of water through the entire river system. This flood would wipe out the weeds, improve water quality and give the river the chance to regulate its health. It would destroy homes, livestock and irrigation infrastructure – yet it is longed for as 'an expression of the freedom, energy and life-giving power of the river. The flood would return us to the times of droughts and floods, when the river was free of dams and weirs. The flood's watery blanket would cover all places and seep in, returning country to the rivers' domain.'

I understand this longing, this impossible dream. I had left the hardest part of my river journey until last.

I did a great deal of my growing up in South Australia's Riverland, in particular in the town of Waikerie, an irrigation oasis in the middle of marginal wheat country. Here I spent most of my girlhood weekends and school holidays. Here I fell in love for the first time. Here I had the epiphanies to which adolescents are prone. My first job was cutting apricots for drying in a corrugated-iron shed by the river. The only bone I ever broke was on my left hand, which got caught in a rope I was using to swing over and into the river. It is still slightly bent. It was in Waikerie that I got my first job as a journalist, working on a newspaper with a circulation of two thousand, the *River News*. And it was here that I returned in the early 1990s, the time of the blue-green algae outbreak, to write my first novel – which was set in a town very like Waikerie.

Most Australians have soul country, a place that lodges in and helps to set identity. There is a particular pain, a particular bitterness and a particular confusion in realising that yours smells of death.

I don't go back to the Riverland very often these days. When I do, it hurts. But last March, I returned for my stepson's wedding. I remember him from when he was a baby. One very hot summer day, before he could walk, I sat him and his sister in big buckets of water watching television, in a house closed against the sun. It was unbearably hot. Yet, unless my memory is faulty, I am sure that the orchards across the river were wreathed in the spray from overhead sprinklers. How we wasted water then, without even knowing it. These days, there are no overhead sprinklers. There is only the careful measure of drip irrigation, and not much of that.

On the drive to the wedding, we passed through the wreckage of orchards where trees had been ripped up and bulldozed into piles. We saw the dead grapevines, and the remaining orchards, hanging on. In the town, the sporting fields and parks were brown under the sun, the council forced into the hard calculation of which areas to keep alive and which to let go.

In the weeks that followed, I rang the office of the South Australian Minister for the Murray, Karlene Maywald, and asked why the state was losing its trees and vines while in Griffith there were still gardens in the desert. 'Because we are downstream,' came the bitter answer.

Just a few weeks ago, Maywald's office issued a celebratory media release. Thanks to rains, South Australian irrigators would be able to access a third of their water entitlements: an increase of 9 per cent on the previous year. That this is a cause of celebration – just a third of the allocation – is a reminder of how desperate things have become.

My stepson's wedding was held in a little limestone church on the rise of the main street. Wedding photos were taken against a backdrop that once would have included a flood plain. Now it is a grey wasteland.

And just a few months later, my stepson and his new wife moved to Melbourne. He had been working at the winery in Berri. Before that, he had worked on the family farm, helping, at times, to grow pumpkins and grapes. He used to deliver pumpkins to the back door of the local supermarket, where he was paid 35 cents a kilogram. He saw the same pumpkins, cut and wrapped, on sale a few hours later at two dollars a kilo.

Now, he had concluded, his future lay elsewhere. There was no future in the Riverland. Even the iconic Berri juice factory is closing down, and my stepson wants to go to university. In the meantime, he has a job in an inner suburban supermarket as a shelf-stacker. He was recently moved from packaged goods to the fresh-produce section. He tells me that this supermarket – one of the big two – throws away a lot of food. He has seen fresh carrots, slightly split, in the dumpster, joined by heads of broccoli, tomatoes that look slightly less than red and sometimes produce that has nothing wrong with it at all, but which has not been bought in sufficient quantities by a picky public.

Staff are not allowed to take the food. As it is thrown away, it is counted to make sure that nothing goes missing. The dumpsters are locked to prevent pilfering. Perhaps it is the sign of a rich country, a country used to feeding itself, that each day all that food – and all the water debt, the effort and the economics it represents – is carried away to landfill.

Margaret Simons chairs the Foundation for Public Interest Journalism, convenes the journalism program at Swinburne University and writes about the media for *Crikey*. Her many books include *The Meeting of the Waters* (Hodder, 2003) and a biography of Malcolm Fraser, written in collaboration with him, to be published by Melbourne University Press in March 2010. Her essays have appeared in *Griffith REVIEW: Addicted to Celebrity, People like Us, Getting Smart, Unintended Consequences* and *Cities on the Edge*. www.margaretsimons.com.au

REPORTAGE

Tulips to Amsterdam

Tasmania as the new food bowl

Natasha Cica

TASMANIA'S Labor Premier, David Bartlett, was sensitive to the critical attention his state has long attracted on the mainland when he addressed the National Press Club in October last year. Challenging the national stereotype that Tasmania is a backwoods basket case, he pointed to an assessment by Commsec which showed that the Tasmanian economy was the best performing in Australia.

Bartlett no doubt surprised the scribes when he painted a picture of twenty-first-century economic prosperity built upon Tasmania's strategic advantages: water, renewable energy and broadband infrastructure. The address was part green paper, part election pre-launch, and the Premier was unapologetically bullish about the future.

Central to his manifesto was the idea of establishing rainy Tasmania as the nation's food bowl. Although Tasmania accounts for less than 1 per cent of Australia's land mass and 2 per cent of its population, the state has 12 per cent of its water. Bartlett identified initiatives in irrigation, innovative agribusiness, low-emission transport systems, and food, tourism and skills development as critical to his food-bowl strategy.

It was an ambitious and astute pitch that Bartlett's down-home political detractors will be hard-pressed to fault. Smart farms with niche fields bursting with produce are the marketing equivalent of Nanna baking organic apple pie. 'We are the only place on earth that exports Fuji apples to Japan and tulips to Amsterdam,' Bartlett boasted. 'We are the southern hemisphere's greatest producer of saffron, a substance literally worth its weight in gold.'

Who wouldn't want more Wild Wasabi Ashgrove cheese – just one of many desirable edibles produced by a northern-Tasmanian business owned and run by a fifth-generation dairy-farming family – plonked on premium platters from Milan to Mumbai? Who couldn't applaud the generation of better educational and employment opportunities along the paddock-to-plate supply chain, against a backdrop of entrenched Tasmanian underperformance on every measure of social and economic security, including the purchase and consumption of healthy food?

PREVIOUS PAGE: *The Task: Supper.* Photographer: Jessica Cassar
from the *Sydney Morning Herald* Shoot the Chef! 2009 photographic competition.

JONATHAN WEST, WHO returned to Hobart from Harvard University in 2006 to head the Australian Innovation Research Centre, is the ideas architect of the food-bowl vision. It is articulated in a new report commissioned by the Tasmanian Government, 'An Innovation Strategy for Tasmania: A New Vision for Economic Development', released in October 2009.

West estimates that an additional five billion dollars can be generated from Tasmania's combined agriculture and food industry by enhancing the wine, dairy and aquaculture sectors. That's equivalent to ten thousand dollars for each Tasmanian. If these targets were realised, it 'would roughly double Tasmania's total out-of-state sales and make Tasmania the richest state, per capita, in Australia'.

West is a former director of the Tasmanian Wilderness Society, and is both passionate and pragmatic about optimal use of the state's land and water resources – and refreshingly free of party-political bias. In developing the strategy, West looked to his intellectual favourites, chiefly Wilhelm Röpke, an economics professor who advised Chancellor Konrad Adenauer on the policies underpinning West Germany's postwar economic revival.

Central to Röpke's approach was the *mittelstand*, the large segment of the economy made up of companies with around a hundred employees. 'The economies of Germany and Switzerland still to a very considerable extent draw on Röpke's philosophy of encouraging the distribution of wealth-producing property,' West explains. 'It matches the needs of Tasmania's food industry. The right structure for our wine industry would be two hundred independent companies, not one giant company with ten thousand employees. The right structure for our dairy industry would be three or four processors, several hundred producers, and three or four Tasmanian-owned and managed dairy companies with Tasmanian brands, competing with one another.

'Places that have this kind of economic structure, not necessarily food-based, are very healthy societies. When you have distributed property – not just housing, which is consumption property, but wealth-producing property – you have a very egalitarian society; you induce very desirable cultural characteristics. People stand on their own two feet; they're independent of government and large bureaucracies and hierarchies; they speak up; they tend to be democratic. It's difficult to maintain a democracy when you have a large number of people who have nothing to sell except their labour, and a small number of people who have vast wealth.'

Throw high-end agriculture and food, along with associated tourism and culture, into the Röpkerian mix and Tasmania – like Provence, Tuscany and the Napa Valley – will be recognised as one of the world's most desirable places to live, a 'real economy based on real production', according to West.

There's many a slip twixt theory's cup and application's lip. But West is confident that Tasmania's food sector can realise its potential, provided government delivers a regulatory framework that facilitates, rather than obstructs, appropriate initiatives.

That's no insignificant proviso. 'If I hear about another national plan, I'll scream,' he says. 'Part of the problem we have in Tasmania is so often our view of things is coloured by Australian taken-for-granteds. We're part of Australia, so we live in a national discussion that is dominated by a physical reality that is very different from ours...we're short of water, but we're not in Tasmania; it's too warm, but it's not in Tasmania; agriculture is a mature and shrinking sector, but in Tasmania it's not mature and it's not shrinking.'

West is certain that establishing Tasmania as the nation's food bowl will depend largely on private initiatives, on a relatively small and fundable scale. 'We've got enough people who are really determined to get on with it, people working in the sector,' he says, 'including investors and entrepreneurs from outside Tasmania who recognise the opportunities and get together with local farmers.'

The food critics Leo Schofield and Matthew Evans relocated from Sydney to rural Tasmania in recent years, and have already spotted their own irresistible opportunities, revamping their own brands with tasty Tasmanian morsels. Welcome arrivals, for sure – but can individual players really drive enough food to market, at the right price, so that Tasmania can spread its stuff on the tables of the nation and beyond without substantial government intervention?

NICK NIKITARIS, A home-grown operator, is no ideologue – 'actually I'm quite right-of-centre in many of my commercial viewpoints' – and believes politicians must take more responsibility for reversing current trends in the sector. The Nikitaris family business is Hill Street Grocer, described as 'gloriously fecund' by Peter Timms in his recent book, *In Search of Hobart* (UNSW

Press, 2009). Located in the newly trendy inner suburb of West Hobart, it's a corner shop perhaps like no other in Australia. Dirt-dusted South Arm pink-eye potatoes nestle near Houston salad greens, Nichols chickens, Tongola goat curd, Grandvewe sheep cheese and Domaine pinot noir; the deli is stuffed with Ziggy's sausages and sweets made by Nikitaris's Greek mother. An Orthodox icon watches over the constant bustle.

Nikitaris and his wife, Natalia Urosevic, left Tasmania after university, to work in banking and finance law in Melbourne and beyond. In 2001 they came home and went into partnership at Hill Street with Nick's brother Marco; a third brother, Nektario, recently opened a spin-off store on Hobart's eastern shore. 'When we came back to Tasmania, we asked: What makes this business popular? Who is the customer, and what do they want? The answer was: an intimate personal relationship, and the knowledge they are purchasing something fresh and locally produced,' Nikitaris says.

'And we find as soon as we offer anything that is locally produced, it will sell out the door, even if it's at a slightly higher price than anything else. Our silver beet comes from Midway Point, the leeks and Dutch carrots from a grower in Devonport, fresh wasabi from a lovely lady in Georgetown, Tina's Herbs from Birchs Bay. We have a cherry farmer who only grows cherries for us, and the rest go to New York for export. We have a view that if we support the local grower they will support us, and it will support a local family, and it will continue to make something better of the local community that we're in. It's not a motherhood statement – it's the reality.'

Nikitaris then points to more awkward realities. 'There are a lot of beautiful things happening in Tasmania at a very small scale, but generally what's happening out there in the world of farmers and small producers is quite a difficult picture at the moment,' he says. Late last year, the fair-pricing fight between Tasmanian dairy farmers and National Foods (with Fonterra and Murray Goulburn, one of three major offshore companies who purchase and process Tasmanian milk) led to a statewide consumer boycott of National Foods' Pura milk.

'Everyone's talking about the fact these milk producers are being suffocated by a ridiculous price, way below the cost of production. Everyone's saying isn't Pura greedy, and in fact they are greedy, and they are in the wrong – but no one's asked why. How did they get themselves in this position in the first place? Pura has two customers, Coles and Woolworths, that represent

90 per cent of the market share in Tasmania. That's the highest in the country, and we also have the highest supermarket grocery prices in the country and the lowest incomes. So what's happening to Pura is happening to a lot of producers across Tasmania.

'There's a terrible consolidation of growers. We used to have lots of cauliflower and iceberg-lettuce growers in the south of the state, for example – they've all now disappeared. Plus, 30 per cent more fruit and vegetables are coming into Tasmania every year since seven-day trading was introduced – this is the fourth year now. So when they talk about paddock-to-plate and that sort of thing, the commercial regulation of the retail side has an enormous impact on how it nourishes small producers, or doesn't. I'd say to the super-markets: Okay, you've got 90 per cent of the market share, but from now on X per cent of all of your production needs to be locally based, and you need to prove that, every quarter. You've taken the thing – now give back. How hard is that?'

EASIER, YOU'D THINK, than spending a hundred million dollars of your own money importing outsized machinery to excavate a huge hole in a sandstone cliff in Hobart's relatively down-at-heel northern suburbs, and engaging Melbourne's Fender Katsalidis Architects to fill it with a supersized waterside showcase for your world-class collection of sex-and-death themed art. That's been the recent focus of the Tasmanian gambling magnate David Walsh, usually described as 'eccentric' and increasingly famous for his forth-coming Museum of Old and New Art (MONA), which will open to the public in late 2010.

Less well known outside Tasmania is his companion foodie initiative on the same site, Moorilla, originally the home of the Italian textile merchant, winemaker and arts patron Claudio Alcorso, who arrived in Tasmania in the 1950s. In 2005, Walsh spent more of his millions building The Ether build-ing at Moorilla, housing The Source restaurant, named after John Olsen's six-metre painting suspended on the ceiling of the building's stairwell. Nearby sits his Moo Brew microbrewery; its boutique beer – 'Not suitable for bogans,' ran the early ads – arrives in champagne-shaped bottles labelled with stylised William Dobell cows. A stone's throw away are the newer MONA accommodation pavilions, featuring furniture by young Tasmanian designers, and bedroom artwork pulled from the rebranded Moorilla wine label, images of wine-soaked naked men and women writhing on pleasure's edge...for the record, Leo Schofield gave the package the thumbs-up.

As a committed vegetarian, Walsh has a keen interest in growing and sourcing quality vegetables, and plans to plant organic hops and herbs at Moorilla, as well as a market garden at his Marion Bay property to supply his restaurant. But his ventures will never be wholesomely Nanna-with-apple-pie, nor Baba-with-baklava. Your afternoon delight in The Source is likely to involve chocolate, eggplant and dark ale. And things may only get wilder when Walsh does a deal with the Sydney-based star chef Tetsuya Wakuda, already a big fan of Tasmanian produce, or another of the 'shit-hot' candidates with whom he's currently in negotiation. (Walsh's words, which mainland newspapers declined to include in the job ad.)

All that's 'true to David and his iconoclastic approach', according to Mark Wilson, Moorilla's business manager, a former chef who trained in Hobart before stints in the late 1980s at La Tour d'Argent in Paris, Claridge's hotel in London and Chewton Glen country-house hotel in the New Forest, and then spent a decade running Tasmanian restaurants. 'He finds the niche in the market that will engage with that sort of thing...we don't have a market drive to cater for the broader community; we welcome the broader community and want it to be accessible, but we know it's not what everyone is looking for.

'I can't speak for David, but for me [our target market is] the politically socially and culturally aware, progressive-thinking people...in Tasmania, and beyond,' Wilson says. Neatly put. As a visionary and revolutionary himself, Professor Röpke would hardly be surprised.

Dr Natasha Cica is the director of the Centre for Applied Philosophy and Ethics at the University of Tasmania. Her book about the Lithuanian-Tasmanian wilderness photographer Olegas Truchanas will be published this year by Queensland University Press. Her essay 'On the ground', in Griffith REVIEW 15: Divided Nation, examined poverty and social exclusion in Tasmania.

From harvest to market

Finding the middle ground

Elaine Reeves

BETWEEN the super- and the farmers' markets is the middle ground – on which growers too small to deal with the supermarkets but with harvests too big to move at a once-a-week market stall find a fit with the new corner stores.

Not so many years ago, a farmer with a few pumpkins or lettuces to spare could take them to the big local supermarket. The fruit-and-veg manager would pay cash for the product and everyone, including shoppers, could sense the relationship between local producers and the local supermarket.

A couple of 'progressive' steps soon put a stop to that. The first was supermarkets creating single distribution hubs – whereby apples grown in the Huon Valley, in southern Tasmania, would be trucked to Devonport, on the island's north coast, before being delivered all the way back to the supermarket in Huonville.

Then, from about 1999, Woolworths, quickly followed by Coles, began insisting that growers who supplied its supermarkets have a Hazard Analysis and Critical Control Points (HACCP) program. This system of checking the danger points in a process, developed by NASA to keep rockets in flight, was to be applied to food safety. In the case of the very small blueberry farm my partner was operating, it meant not merely sweeping the floor or checking the thermostat on the cool store, but ticking a box on the Good Manufacturing Practice Audit Checklist to say you had done so.

It was expensive. Initially, Woolworths subsidised a great deal of the training needed to set up the quality-assurance program, but annual auditing cost about a hundred dollars an hour, and even for the smallest grower it took about three hours. And the grower had to pay a share of the travel and accommodation costs of an auditor coming from the mainland.

In 1992, Phillip Andreatta, a grape grower who had been supplying Woolworths for twelve years, found his price being forced down: the supermarket chain would pay him only $22 for a box of grapes that was fetching eight dollars more at Flemington Markets. Andreatta could not simply stop supplying Woolworths – he had invested so much in Woolworths' quality-assurance scheme, he told the *Sydney Morning Herald* at the time.

PREVIOUS PAGE: *Dinner for two with Manu.* Photographer: Jonathan May
from the *Sydney Morning Herald* Shoot the Chef! 2009 photographic competition.

In our case, after several years of paying quality-assurance expenses equal to about a third of the value of the fruit Woolworths was buying from us, we simply gave the program away. Our practices remained the same, but we no longer ticked the boxes – and we no longer supplied Woolworths.

Many other small-to-medium growers, who previously may have sold to their local big-chain supermarket, either did not start on the HACCP regimen or discarded it after a few years.

IN 2008, I was chasing a story about why a glut of cauliflowers that was wrecking the budgets of farmers was not translating into lots of cheap cauliflower cheese for consumers. Mike Badcock of Forth, in northern Tasmania, who grows caulis for the fresh market on a large scale, explained that growers allowed twelve weeks to grow a cauli in summer and twenty-six weeks in winter. A series of July days at a balmy 15 to 16 degrees 'fooled them into ripening faster', he said. The fresh market was paying Badcock half as much per unit as it had at the same time the previous year.

Richard Bovill was a buyer of fresh produce for Woolworths before he led the Fair Dinkum Food Campaign in 2005, when 170 tractors, mostly from Tasmania, rolled into Canberra demanding support for Australian vegetable growers. In the middle of the cauli glut, I asked Bovill where the specials on the vegetable were. He said caulis are not suited to being on special – a national chain needs a long lead time to organise a special, which has to be available nationwide to suit advertising coverage, and cauliflowers need to be picked within three days of maturing. In the days when it was possible to have a state or even more localised special, supermarkets could take advantage of a cauli glut. Then, cauliflowers were cut in the field and fifty or sixty placed in bins that were delivered to the supermarket.

Now, quality-assurance programs specify that cauliflowers must be packed in waxed cardboard cartons, adding the best part of forty cents to the cost of each cauliflower; the leaves have to be trimmed, which cannot be done in the field, so a packing shed and more staff are required. Handling systems and compliance regimens 'trying to take cost out of the system for the retailers have almost doubled the price of cauliflowers', Bovill said.

'In a world where we look at efficiency and greenhouse-friendliness, we've gone and loaded ourselves up with waxed cartons and layers of stuff that in a lot of cases has made fresh produce much, much dearer than it used to be. It is the reason smaller shops and wholesalers are often cheaper than the big players – because they can still buy produce the old way.'

The major supermarkets prefer to deal only with larger growers who can supply greater volumes, year-round. I know of one Tasmanian tomato grower who would prefer not to cultivate them in hothouses in winter, but must if he is to maintain supermarket access for the tomatoes grown in summer.

THE TRACTORS ROLLED out for the Fair Dinkum Food Campaign belonged to farmers who grow for the frozen-processed-vegetables market, which is by far the biggest horticulture sector in Tasmania. The supermarkets' constraints leave those producers operating on a small but commercial scale looking for other outlets, especially farmers' markets. There are about two hundred registered with the Australian Farmers' Market Association. They are credited with saviour status for niche producers, but this is not really the case.

Our farm is a handful of hectares, yet we grow between six and fourteen tonnes of blueberries a year – far too many, even in the leanest year, to move from a stall in a farmers' market, whatever the attraction of making a retail, rather than wholesale, price. We also grow strawberries and garlic, but there would not be more than a couple of weekends in a year when we could offer even this much variety.

I talked to an organic grower of potatoes who sells two-thirds of his annual crop on the conventional market at conventional prices, because potatoes are oversupplied in organic markets. I asked him if he ever sold his potatoes at the nearest farmers' market, two hours' drive away. 'It's a day getting ready, a day at the market and a day unpacking – it's just not viable in terms of wages,' he said.

Some farmers' markets are less pure than others about insisting that the stallholder has grown or made the produce – but assuming everyone is playing fair, a farmers' market suits a particular type of grower, and it is not the farmer specialising in large quantities of just a few different crops. John Russell brings a truckload of vegetables to the Burnie Farmers' Market on the first, third and fifth Saturdays of the month. He puts up no price signs, relying just on his Con the Fruiterer spiel, and always he sells out. But his farm near Railton also supplies his own fruit-and-veg shops in Launceston and Railton.

Brett and Simone Connors go to Burnie and Devonport farmers' markets on alternate weekends. They grow 'just about everything that's on the table', which includes cauliflower, cabbage, pumpkin and potato, and buy in capsicums and strawberries from neighbours. But they do not rely on the market for their income; they also raise cattle and sheep at their farm at Lower Barrington.

SAINSBURY'S SUPERMARKETS IN Britain sometimes turn their car parks over to farmers' markets, something Australian supermarkets have not yet emulated, but Delicacy, a small food shop in Launceston, holds a farmers' market in its car park ever other Sunday. The shop is far busier on market mornings than on the days when there's no market, and the synergy works for market stallholders.

Frank Archer told me most of his family's Landfall branded lamb and beef was sold through Delicacy, so in June 2008, they began selling at the farmers' market as well. In six months, sales of lamb increased five-fold and they were selling twelve times as many sausages.

It is a rare thing for a producer to be wholly reliant on a farmers' market for their income. Bruce and Clare Jackson run Yorktown Organics, a market garden near the mouth of the Tamar River, in northern Tasmania. The farm produces a hundred lines, which may be different versions of a single vegetable. Beetroot, for instance, is sold as sprouts, as baby leaves, as baby beets and as mature beetroot.

Bruce Jackson's entire farm is one hectare outside and 0.3 of a hectare under glass. He explained, 'We grow high-value product – leafy stuff – tat soy, wild rocket, tomatoes, soft fruit, apples, new potatoes. What we do is very diverse; we're not relying on one thing, not reliant on *a* crop. It would be very hard for a guy who can't see the fence at the other end of his fifty-acre paddock to really understand what I am on about. It's a different mindset.'

Jackson doesn't sell to major supermarkets, nor does he stand behind a stall at a farmers' market. He has a farm-gate shop, and sells directly to restaurants and smaller retailers. He says selling his produce this way prevents him from being forced into lowering prices: unlike the process growers, he is the price-setter.

FORTUNATELY FOR PROFESSIONAL Tasmanian farmers who fall between the super- and the farmers' markets, there are many small, high-end retail outlets in the state for their produce – retailers who have actually been helped by the practices of the big players that exclude or deter small and medium-sized growers. John Price was the owner of the Dover store, in Tasmania's far south, when supermarkets opened in the state in the 1960s. He acted 'boldly' to meet the challenge: out went the gelignite, drapery, shelves to the ceiling and letting purchases mount up on credit (about 95 per cent of customers paid their bills only once a month). In came trolleys and baskets, shoulder-high shelves and pay-as-you-go.

To meet this latest wave of supermarket domination, smaller stores have concentrated on quality goods, their own cooked fare, a good deli. They carry your bags to the car and know your name – and they buy locally from suppliers and growers who do not produce enough for the big supermarkets.

Among the earlier adapters was Hill Street Grocer, formerly a corner shop in West Hobart, which a decade ago morphed into a high-end food store specialising in fresh produce. There is only a small space devoted to fresh produce, but the turnover is much greater than the area suggests, and three or four people are employed to constantly replenish the display. In summer, Hill Street boasts that more than three-quarters of its fresh produce is Tasmanian.

Another type of retailer helping small producers are the hunters and collectors, such as Jonathan Cooper of the All Organic Farm, whose main outlet is a stall at Salamanca Market on a Saturday. Cooper doesn't have a farm, but collects produce from a couple of dozen local organic producers, as well as mainland growers and organic wholesalers, and sells it on. It means farmers are not giving up precious time to stand behind a market stall, and customers have a wide range of produce to choose from.

The gut feeling of these growers and retailers occupying the middle ground was not the same as the conclusion reached by the Australian Competition & Consumer Commission's inquiry into the competitiveness of retail prices for standard groceries. The inquiry focused on whether a lack of competition was the reason for concerns that the gap between what's paid to farmers and what's paid at the checkout was widening at the expense of the farmers, and whether it threatened the viability of small, family-run supermarkets and small greengrocers, butchers, bakers and delis.

Rather than finding excessive mark-ups on goods, the ACCC blamed increases in food prices on drought, weather catastrophes, an international commodities boom, and increases in petrol and fertiliser costs. It found that only a third of Australians did all their meat and fish shopping at the supermarket – a third went to some smaller shops as well as the supermarket, and a quarter did all their shopping at smaller shops.

Research commissioned by Meat & Livestock Australia and presented to the inquiry found that in the past seven years there had been a marked increase in the amount of money Australians spent on fresh food – up 42 per cent for meat, 38 per cent for fruit and 37 per cent for vegetables. This was not because prices for fresh food had shot up, but because more people were cooking at home and looking for seasonal food produced locally. The research found much of the additional spend went to independent retailers.

The federal government's immediate response to the ACCC inquiry was to set up the website GroceryChoice, which was to compare the price of a basket of groceries at major supermarkets and independents. Then the consumer organisation Choice was to take over the website, and provide a more useful service: a consumer would be able to bring up a couple of supermarkets in their area, enter their shopping list and compare the prices. Just six days before the Choice website was to be launched, the Minister for Competition Policy and Consumer Affairs, Craig Emerson, pulled the plug on the project. In a submission to the Senate Economics References Committee Inquiry into the GroceryChoice website, Choice said it still was 'not clear why the decision to terminate the project was taken and firmly believes that it was the wrong decision'. It said some major supermarkets did not co-operate, but also the 'lack of political will to seek legislative and non-legislative solutions had a detrimental impact on the ability of

Choice to deliver the website the public wanted'. This leaves the introduction of unit pricing as the only outcome of the ACCC inquiry.

In southern Tasmania, independent stores fared poorly in the ACCC survey, and were consistently more expensive than Coles or Woolworths, the only majors in the island state. But to get on to GroceryChoice's radar, an independent had to have more than a hundred square metres of floor space. In Western Australia, independents of this size make up a third of supermarkets; in Tasmania, there are only two – 5 per cent of the market – and the we-try-harder small retailers did not make the cut.

Another survey, PriceWatch, comes out of Labor MP Duncan Kerr's Hobart electorate of Denison. It compares a basket of specified items at specified Woolworths and Coles stores in the electorate, and in 2007, for the first time, surveyed small independents as well. It found the smaller shops were cheaper for five of eight fresh items, and overall – figures that 'completely dispel the myth that shopping at small grocers is always more expensive', said Kerr's office. A year later the PriceWatch survey of twenty-two stores showed a difference of $18 between the cheapest and most expensive baskets – both extremes coming from independents.

What's more, choosing the family-run supermarkets, greengrocers, butchers, bakers and delis ranged between the super- and micro-retailers also supports the growers and producers who occupy the middle ground between the twenty-hectare paddock and a garden.

FOR US, THE ability to sell to smaller, specialised shops, and to the market stalls that collect from a number of growers, has meant we sell our entire crop, at good prices, without needing to tick boxes and pay someone to make sure we've done so under a quality-assurance program – nor to take time out in the middle of managing pickers and packers during a busy harvest to set up shop at a farmers' market.

Elaine Reeves is a journalist working in Tasmania. For eighteen years she has written a weekly column on food for Hobart's *Mercury*.

Backyard gardens

Creating future cities

Brendan Gleeson

WE rarely talk sensibly about food – the throwaway line 'food for thought' is about as close as we come to connecting thinking with eating. This isn't an attitude we can afford any longer. Our food production and distribution systems, and all the cultural practices that go with them, are facing some deep challenges. It's time we had much more thought for food.

Lots of things are going haywire. Ocean acidification and resource exploitation are ruining global fish stocks. Rapid urbanisation in the developing world is sucking the life from agricultural regions. Declining crop yields have large parts of the world on famine watch. And, idiotically, we're switching crops from food to biofuel production at a time of soaring populations.

Things are clearly out of whack in our food systems. And there's a much bigger whack that threatens to knock our entire food regime sideways. 'Global warming' is perhaps not the best way to describe the threat. The renowned British environmentalist James Lovelock prefers 'global heating' to describe the human-fired barbeque that is cooking our world.

Global heating is a diabolical challenge that undermines human food supply at a number of levels. First, it reduces that most vital precondition for food, water. Already in Australia, science confirms a permanent loss of catchment capacity in some regions. Worryingly, these include highly urbanised parts, including Perth and the hinterlands, and the populous south-east corner of the continent.

It also attacks food on other fronts: making land less arable; making local climates less predictable and thus less productive; and reducing biodiversity and future potential food sources. A recent international study predicted that if heating continues unchecked, global wheat production will drop by up to 27 per cent, and rice by 13 per cent, over the next forty years.

The climate slow-burn will steadily erode the human resources for food production. It is stressing Australia's inland rural communities, where interminable drought manifests in soaring suicide rates. A journalist from the property pages of a major daily told me a few years ago that inland farmers were quietly buying up land nearer the coasts, expecting to need to shift from untenable farms. Others will simply walk when things become intolerable.

In September 2009, the CSIRO chief, Dr Megan Clark, told an astonished National Press Club that 'in the next fifty years, we will need to produce as much food as we have ever produced in the entire human history.' This, at a time of declining agricultural possibilities, much of it attributable to climate change.

A bleak assessment, but the optimists, or at least one known for his noisy buoyancy, had another view. Senator Bill Heffernan urged us to look to Australia's north as a new frontier of agricultural ambition. Among conservatives, Heffernan is notable for taking warming seriously. His cheerful antidote proved hugely newsworthy. The CSIRO doubts northern Australia's suitability for food production. But these and other expert objections were swept aside. Science, the senator was certain, could unlock the secrets of tropical productivity that have eluded us for a very long time.

WE'LL CERTAINLY NEED a great deal of human ingenuity and a good resource base to draw upon in the battle against global warming. I believe that ingenuity is much more significant than science − critical as formal expertise is, and will be − and that we have a resource base much closer to hand than the far north which is much more suitable for development in quest of resilience.

I refer to the ground you are probably sitting on as you read this, suburban Australia. This is where the overwhelming majority of Australians live and will do so for a very long time, whatever our urban redevelopment ambitions. The 'fibro frontier', the 'veneer frontier': it has many manifestations, but shares a few powerful and useful qualities that we should consider.

First, the vast suburbia of our metropolises and sea-change regions occupies some of the best watered and most productively 'soiled' land we have. And climate projections suggest that they will continue to receive acceptable amounts of rainfall. Second, with the exception of more recent 'small lot' estates, suburbia is a low-density greenscape with a lot of disorganised but potentially productive land.

There are obvious barriers to producing food in suburbia – it's a fragmented, privately owned patchwork, for a start. There's also the tricky problem of protected pests that thrive in our green suburbs: possums, crows and the like. The bush turkey that regularly visits our Brisbane garden loves to uproot anything edible. Possums eat our potted chilli bush with unfathomable (and happily unseen) results. But perhaps it's time to think of adding that other ingredient, human ingenuity, to the mixture of resource use. A reservoir of quiet innovation exists in suburbia. We discount it at our peril.

The suburbs are often wrongly miscast as anti-environmental, which is ironic, as Australian environmentalism was conceived and hatched there. There is plenty of evidence that the environmental sensibility remains, slumbering perhaps, but ready like all sleepers to be awakened by the right cause. Think of the stunningly effective response of suburbanites in South-East Queensland during the terrible drought of 2000–07, which took our third-largest conurbation to the edge of possibility. With resolute state and municipal leadership, householders were able to reduce per-capita water use to the lowest levels in the developed world, and crisis was averted. As my Griffith University colleague Geoff Woolcock points out, the suburbs rescued the situation. We may now ponder the legacy of this and other responses to urban water crises in Australia. Many residential properties have adapted permanently to the need for water conservation. Tens of thousands (more?) of households now have independent access to on-site tank water. This must amass to a substantial new catchment capacity in the cities. Presently it's used to keep lawns green and roses blooming, but it could easily service a new suburban food endeavour.

We've been there before. In fact, until relatively recently, our suburbs were highly productive food regions. In simpler times, the dictate of self-sufficiency was carefully maintained. The Queensland sociologist Patrick Mullins demonstrated that even up until the 1960s, a very substantial proportion of produce consumed in the cities came from suburban backyards: chooks, vegies, fruit that became preserves and jam. And the rest was mostly sourced from the immediate hinterlands of the cities.

In *Harvest of the Suburbs* (UWA Publishing, 2006), Andrea Gaynor documents the long history of suburban self-sufficiency. Labour was often divided along gender lines: dad tended the vegie plot and the fruit trees, and mum the chooks. In times of stress — such as war or depression — the suburban soil was tilled harder, and with great success. We were not unique. James Lovelock recalls that in Britain during World War II, 'a great surprise…was the discovery that the output of food per acre was four times greater in gardens and on allotments than it was on farms'.

Another argument in favour of suburban agriculture is the need to de-carbonise food. We need to radically reduce food miles to reduce the energy used by the economy. We must, in short, stop burning the food before it reaches our mouths. A low-carb(on) diet must become the norm, not a fad.

The suburbs beckon a new, comprehensive makeover which will make them fit for food production. This means dispensing with those unbending critiques of suburbia which neglect its vast latent potential to aid climate adaptation and social resilience. The CSIRO's Ian Holmgren writes, '"Suburban sprawl" in fact gives us an advantage. Detached houses are easy to retrofit, and the space around them allows for solar access and space for food production. A water supply is already in place, our pampered, unproductive ornamental gardens have fertile soils and ready access to nutrients, and we live in ideal areas with mild climates, access to the sea, the city and inland country.'

Yet leadership is lacking. There is little discussion about cities, other than their role as the sites of 'shovel ready' infrastructure projects. The states, which carry principal responsibility for urban management, have snookered themselves with inflexible visions of the compact city, freighted with much impotent anxiety about the 'sprawling' suburbs overwhelming the food bowls that once marked the edges of the cities. At the municipal level, things are more promising but also patchy. Community gardens are flourishing, yet are unlikely to become a major source of food supply.

We need new metropolitan commissions to independently manage the cities, and guide us through the unfolding climate and resource crises. Such a model would be far removed from the venality and occasional villainy of state politics. It would develop a coherent approach to urban food production.

THERE ARE OTHER barriers to suburban agriculture that we must ponder. Given the new, diverse social structure of contemporary suburbia, we cannot simply return to the former system. Women are at work, not minding kids and chooks, and we all work more intensively these days. Who will tend the plots?

Perhaps part of the answer is to end our hand-wringing about population ageing and recognise the grey, and largely fit, army that might be willing to undertake a new type of gardening. Equally, food production in schools, as Stephanie Alexander has advocated – potentially tens of thousands of school gardens – would provide a new focus for education and involve children in nurturing and consuming healthy food.

There is a host of pesky sundries that have to be addressed: private-property rights, public-liability issues, health and safety concerns relating to non-organic produce, safe storage of water and the like. These details will bedevil us unless they are addressed through legislation, help for householders and, above all, municipal guidance. Councils should be the everyday managers of suburban food production, and provide community exchanges for the sale and trading of produce.

In generations past, quiet necessity brought the 'harvest of the suburbs'. It ended with our brief flirtation with fantasy, when we believed ourselves utterly freed from Nature and released to the freeways of bound-less gratification. A fire in the heavens now glowers over us. This is the real barbecue-stopper of our times.

We've tried running away from Nature for a long time, and look where it got us. It's time to come home and tidy up our own backyards. I'm certain that we'll be surprised and reassured by what's possible.

Brendan Gleeson is a professor of urban management and planning, and the director of the Urban Research Program at Griffith University. His essays have been published in Griffith REVIEW: Cities on the Edge, Re-imagining Australia and Hot Air.

Creating sustainably productive cities

From permaculture to urbaculture

Virginia Balfour

THE schoolboys laughed when they were told apples grew on trees. 'Well, where do you think they come from?' I asked. Looking at me as though I was from another planet, one replied slowly, 'From the supermarket, of course, Miss.'

It is a funny story, but also a tragic one. Do many of us really no longer know where our food comes from, before it fills bags in Woolies or Coles?

For a long time, it might not have mattered. But as we face the challenges of climate change, spiralling population growth and serious resource depletion, there is a need to reconnect with what we eat and where it comes from.

In 1972, a group of scientists rather pretentiously calling themselves the Club of Rome published *The Limits to Growth*, in which they predicted that if the world continued to grow and consume at the same rate, we would face social and economic collapse within a century. There was much debate about the report at the time, but only in 2008 were its predictions tested. The CSIRO scientist Graham Turner compared them with data on what had actually happened to consumption levels and resource depletion. He concluded that we have continued on a 'business as usual' pathway and, almost forty years down the line, the Club of Rome's predictions are still the most likely scenario. We have reached a point of crisis. Climate change, as Prime Minister Kevin Rudd said in his concluding statement at the 2020 Summit, is 'the overarching issue this generation and those to follow must address'.

Richard Eckersley, a visiting fellow at the Australian National University, says our reaction to a crisis can manifest itself in one of three ways: nihilism (it's all hopeless so let's enjoy ourselves while we can); fundamentalism (falling back on the likes of religion or the free market); activism (there is hope if we apply ourselves to the problem with sufficient urgency).

The Transitions Network may be a starting point for activism. Based on a concept developed by Rob Hopkins in Ireland, five years ago, Transitions revolves around the idea that the combination of climate change and peak oil will mean we have to radically change the way we live in wealthy countries.

PREVIOUS PAGE: *Cooking up a storm (detail)*. Photographer: Cassandra French from the *Sydney Morning Herald* Shoot the Chef! 2009 photographic competition.

The Transitions Town Initiative is 'a catalyst for community-led responses' and sets out strategies for 'regions, counties, cities or even neighbourhoods' to formulate a plan and retrain themselves to live in an oil-free, carbon-neutral society. Hopkins describes it as an 'extraordinary opportunity to reinvent, rethink and rebuild the world around us'.

At a Transitions Network East Brisbane meeting, fifty people gathered in a local park to discuss *The Age of Stupid*, a film about climate change, and formulate a collective response to deal with the issues it raised. The man beside me came up with four words to describe how the documentary made him feel: present, clear, focused, determined. Those words seemed to sum up the mood. Alongside lobbying and creating local networks, one of the key ideas proposed was local food production, something *The Transitions Handbook* describes as 'the most sensible place to start'.

Rob Cross and Roger Spencer, in their book *Sustainable Gardens* (CSIRO Publishing, 2008), suggest one of the key messages is to eat less meat. This view was echoed in the Stern report, which pointed out that our attitude to eating meat is going to be one of the most important elements (alongside energy and transport) in efforts to cut greenhouse-gas emissions. In Australia, we produce and eat an awful lot of meat. As the author and animal-liberation campaigner Geoff Russell, citing recent CSIRO research, points out, Australia's meat industry generates more greenhouse gases than the transport sector. Tim Flannery says, 'we should be eating what is good for the planet, as well as what is good for us' – and that is definitely a lot less meat. And if we are to eat less meat, we should be considering eating more vegetables, grown in a sustainable way.

I MENTION TO a neighbour that I'm thinking of replacing the strip of lawn outside my house with tomatoes and rosemary – a sort of nature-strip tease, if you like. 'Oh, you can't do that!' she says in horror. 'It's illegal.' Now, I'm not an anarchist, but when someone tells me that I can't do what I want with a piece of land outside my front door, I get a bit indignant. I protest, but am told in no uncertain terms that, although I must look after it, that patch of grass is council land and the council wants it kept as grass.

Another neighbour sidles up to me and whispers, 'Do you want to buy some eggs?' His tone is conspiratorial, as though he is trying to sell me an illicit drug. 'Of course, if anyone ever asks, you'll have to say you didn't pay for them.' He winks. 'Just say you were given them and you paid a donation for the chickens' upkeep.' It seems that council policy doesn't allow locals to sell their eggs to each other. You can give them away, but you can't sell them.

This is part of the problem. We need to create a green revolution within our urban areas, promoting the idea of organic-vegetable growing and urban agriculture, but bureaucracy is thwarting us. Brisbane City Council has a vision for creating 'Food in the City', promoting 'healthy and active lifestyles' and providing grants for new community gardens across the city. And it is developing a new policy on street trees; they can be productive food trees, so long as your neighbours agree and the trees don't cause a mess or a pest hazard. It's a start, but it is still a way behind cities in other parts of the world.

In San Francisco, the mayor's office called in mid-2009 for all council departments to audit land suitable for, or used for, food-producing gardens. The city has established a Food Policy Council, incorporating everyone from the mayor's office to the parks department, local restaurants to the retail food industry. The Department of Recreation and Parks will co-ordinate urban agriculture, including access to gardening materials and tools, and food-production and horticulture education is being undertaken within the city, especially on city-owned land.

SOME GREAT THINGS are happening here with our kids, at least. My local primary school, Bulimba State School, is the Queensland showcase for the Stephanie Alexander Kitchen Garden Program. The program, started in Victoria, teaches primary-school students how to grow and use vegetables. In 2009 it went nationwide, starting in thirty-seven schools around Australia and attracting $12.8 million in federal assistance. Each week schoolchildren spend at least forty-five minutes maintaining a vegetable garden they helped to design and build on the school grounds. The program revolves around growing and harvesting, preparing and sharing, and aims to provide enjoyable food education to young people.

Sara Breckenridge is working with prep and grade-one kids at the Bulimba school. She has overseen the conversion of a central area of the school to a stepped garden growing everything from broccoli to sunflowers. There is even a school scarecrow. Everywhere you turn in the grounds, there seems to be a pot in which something is growing. There are plans to establish a bush-tucker garden and convert more of the grounds to vegetable production.

I went along in late 2009 to help make beetroot dip. We picked, washed and grated the beetroot, mixed it with garlic and natural yoghurt, and ate it with bread. Yes, I did need to pinch myself when I saw a roomful of five-year-olds polishing off a bowl of beetroot without a peep of complaint. A mother told me that her son loves the garden, and can't wait to get out and work in it. These are kids who know where apples come from.

What about those mums who turn up at the school gates in truck-like cars: will they be persuaded by their eco-kids to start growing their own? One mum I know sat on the edge of her kid's sandpit at home and triumphantly told me she'd ripped out all the vegetables put in by a previous owner, because they were 'too much trouble'. It's an understandable attitude given how far we are from the days when we all grew our own.

Australians have a long history of growing their own vegetables. Home gardens and Victory gardens were a necessity during the Depression and world wars. My friend Pat is typical of that generation. She has grown vegetables all her life, and although elderly now, she still has a little bit of the backyard to grow a few plants: a pawpaw, tomatoes, some greens. In my letterbox the other week, there was a small envelope with seeds from Pat's rosellas, along with a handwritten guide to planting them. My daughter and I had spent an evening with Pat months earlier, tearing the leaves off her last crop and boiling them up to make delicious rosella jam, and now she was giving us some seeds to grow our own crop. This is how things used to be done: a neighbour passing seeds or cuttings on to a neighbour, friends exchanging excess produce, each connecting with the land and the seasons and helping the other along.

Today, we seem to have lost that knowledge. The president of the local senior citizens' group told me that one of her neighbours rested a pin-striped arm on her gatepost on the way back from the office and asked what the lovely plant was in her patio garden: 'The one with the green leaves and the red flowers.' 'You mean the geranium,' she replied. She told the story while rolling her eyes skyward. 'Fancy not knowing what a geranium was – it's probably the most popular flower in Australia.'

IF WE ARE to tackle climate change and make a difference at a community level, we need to educate all Australians about how to live sustainably. We will need to educate people who drive cars and have mortgages and jobs, and all the pressures that go with them. We need to learn how to live locally, to eat less meat, and grow our own fruit and vegetables a lot nearer to home than we do at present.

So here's my idea: urbaculture. It's based on permaculture – a neologism formed from the words 'permanent' and 'agriculture' – which is the system for sustainable agriculture developed in the 1970s by two inspirational Australians, Bill Mollison and David Holmgren. Urbaculture is permaculture for the city dweller. It is a blueprint for sustainable, food-producing cityscapes, where food is grown locally and organically within communities and suburbs.

One of the first steps to establishing such a system will be thinking about how we bring food production and food education in to our cities.

Could we use our public parks? Brisbane's Botanic Gardens were originally set up as experimental gardens, trialling crops and plants to determine their suitability for the local climate. They were used to feed the colony. Could we do the same with bits of our public parks today?

Every suburb of our cities has a park: a large open space, bordered by trees and carpeted in grass. There is often a small playground in one corner, and on Saturday mornings children might be seen playing soccer and fathers frying sausages. During summer, the council sends along a big tractor to mow the grass, but most of the time most of the park sits there unused and baking in the sun.

Inner-city public parks date back to Victorian Britain, when governments were keen to provide diversion and healthy exercise to the workers from local factories. In far-flung parts of the Empire, they often included a tearoom and rose gardens to remind the expats of home. The model no longer serves the needs of either urbanites or our environment.

Some of our parkland could be used for productive food growing, or as education gardens where we teach communities how to grow organic vegetables. Not all of it, of course – we need to keep some space to kick a ball and have a barbie – but there are unused corners of our parks and parts too exposed to be enjoyed.

My local park is Bulimba Memorial Park. It was originally swampland, and the journey to Bulimba from the city was an arduous series of boat trips around the swamp. Just over a century ago, the swamp was filled in and made in to a park. After World War I, the park was designated a Memorial Park and eighty trees were planted around the perimeter, each one commemorating a soldier or sailor. The park was renowned for its cricket oval and grandstand. Over the years, there have been some modest changes, but ostensibly it has stayed the same for close to a hundred years.

I have proposed that we develop a community-education garden in the park. I have the backing of a lot of local residents, businesses and community organisations, as well as local politicians, but it's a heritage park and the council is, so far, reluctant to change. One day, however, I hope it will be seen as a model for all our public gardens: as parks for people, parks for pleasure, but also as parks for food production and education.

Some people may wonder whether all this is really necessary. But as Richard Branson says, 'I'd rather be an optimist and proved wrong than a pessimist and proved right.' Growing our own food within our cities may be small beer compared with the radical things we are going to need to do to address climate change and the likelihood of peak oil, but it is something we can do now – and do relatively easily. And when someone asks where apples come from – where they *really* come from – our kids will know the answer.

Virginia Balfour is a journalist and former television producer trained in horticulture, permaculture and international environmental policy.

Feeding the world

Our great myth

Cameron Muir

NEVILLE Simpson is not your typical cotton farmer. He doesn't hold a university degree, nor does he command tens of thousands of hectares. He doesn't have time for cotton-industry PR, and he doesn't talk fast. He's not American or British, and neither is his business. He lives where he farms, on the Darling River near Bourke, and this alone tends to set him apart. He's elderly, softly spoken, with a slight western drawl, and takes any opportunity to make a self-deprecating aside about his farming expertise. He's reflective, not reactionary, and this is probably why the good journalists often find their way to him when they report from a town that has long represented the quintessential 'rural' locale in the Australian popular imagination.

I met Neville two years ago, when a colleague and I drove to Bourke to undertake research in the library's local-history collection. We'd also planned to fit in some time walking the floodplains, as well as hiking and camping at Mt Gundabooka National Park. Over the past two years, I have been travelling the Macquarie Marshes and Barwon-Darling country, researching the history of agricultural and conservation science. Speaking to people like Neville, and wandering through the country, offers the chance for a better understanding of rural place, agriculture and the environment than could ever be achieved by listening to headline-grabbing lobbyists, politicians and agro-industrial input suppliers – those who say they speak for and want to improve the lot of farmers and rural communities, but who are motivated by a desire for votes and money, and who are working on a national or international scale, not an ecological or bioregional one.

Take the National Party's John Cobb, the Shadow Minister for Agriculture. 'Why has the Rudd Government declared war on the town of Bourke?' he asked, after the British-owned Clyde Agriculture sold the iconic Toorale Station to the federal and New South Wales governments in 2008. Toorale, at the junction of the Warrego and Darling rivers, is conspicuous for its squattocracy and corporate-farming heritage, as well as for being the place where Henry Lawson spent a month working as a rouseabout in 1892. The federal

---

PREVIOUS PAGE: *Chef Pierre Labancz (detail)*. Photographer: Chris Searl from the *Sydney Morning Herald* Shoot the Chef! 2009 photographic competition.

government has taken Toorale's water entitlements, and NSW National Parks is managing the 91,000 hectares of land. According to Cobb, the outback community of Bourke, sixty kilometres upstream from Toorale, was 'sacrificed to the environmental green-shoe brigade' by a government that 'couldn't care less about the future of regional Australia'.

Bourke shire councillors, vying for re-election in the weeks following the purchase, staged a rally protesting against the government's actions. Their scaremongering spread fear in a community already doing it tough. The *Sydney Morning Herald* reported that the local newsagent was 'petrified' for the future of the town. John Cobb described the purchase as an 'anti-rural-Australia act'. Inevitably, a well-worn political catch-cry appeared in media comments: if the government was putting the rural sector out of production, instead of supporting it, how could Australian farmers continue to 'feed the world'?

THE GOAL OF feeding the world is an admirable one, but it does not reflect the reality of Australian farming. Most of Australia's wheat and meat are exported, and this has become the basis for a national myth, a comforting narrative that sees golden harvests and choice cuts being shipped and distributed to hungry mouths across the world. In 1925, the leading Australian meat-industry figure John Cramsie declared that the development of the 'unoccupied northern areas' presented an opportunity to 'feed the world with beef'. The prominent doctor and journalist Edward Gault gave an address in 1943 arguing that Australia should not only feed India and China, but 'it should be a permanent measure for us to feed the world as a whole.' After the sale of Toorale Station in 2008, the Nationals' leader, Warren Truss, told parliament: 'We cannot keep taking properties out of production and expect to meet our obligations to provide food to the world.' Paul Myers, a former editor of *The Land*, wrote an editorial about agriculture and Toorale for the *Sydney Morning Herald*, asserting that Australian farmers 'contribute significantly to global nutrition'.

In fact, Australia contributes less than 2 per cent of global food production. In 2004, Australian-grown wheat and other cereals accounted for just 1.39 per cent of the world's grain production. It is a similar story for Australian meat, which provides 1.45 per cent of the world's share, and for fruit and vegetables: a tiny 0.4 per cent. These United Nations Food and Agriculture Organization (FAO) statistics have been consistent for the past quarter of a century. Consider, too, that most of our food exports are premium products, and we sell nearly half of all our export produce to just five rich nations: Japan, South Korea, the UK, New Zealand and the USA.

Why have rural lobbyists talked about feeding the world when it is clear that Australian agriculture makes a very minor contribution to world food production, and hence has no significant role in nourishing the world's poor? Because, throughout the history of settler Australia, we have valued the social function of agriculture over its utilitarian function. We have cared more about the culture, character and work of agricultural production than the actual food and fibre it produces. That social function is being reassessed, and the role of agriculture in Australia might change – again.

TOWARDS THE END of the nineteenth century, the settler project was in a dire state. 'The promised land was about to forget its promise,' Geoffrey Blainey wrote in *A Land Half Won* (Sun Books, 1983). The pastoral industry had collapsed and economies sank into depression. Inland villages, in the words of one writer at the time, were left 'sun-devoured and sand-swept'. The colonies began losing their populations through emigration. Whenever the coast-hugging settlers turned their gaze toward the great interior plains, they glimpsed broken country, bloodshed and extinction. They saw skulls pierced with blunt lead bullets, ribcages cracked open with heavy spears, red country littered with ringbarked timber and the desiccated carcasses of millions of sheep, claypans, silted creeks and sagging slab huts; they saw the material remains of initial hopes and land lust bleached by an unrelenting sun. For governments, the frontier quickly became a liability; for pioneering colonisers, it was a path to ruin.

Governments needed to redeem the settler project, and agriculture based on scientific principles emerged as an unexpected solution. The idea that agriculture could play a significant role in the development of the colonies had been largely abandoned as settlers focused on mining and pastoral pursuits.

Few people thought agriculture could pay in such a dry climate, with such a small population and long carting distances. In 1892, William Wilkins, the Under Secretary for Public Instruction, wrote a fifty-page treatise on agriculture in New South Wales, which began, 'It was a maxim of ancient statecraft that the food supply of the people should be raised within its own boundaries.' Wilkins went on to comment, however, that imperial relationships and international trade had rendered this obsolete. New South Wales, following Britain, could import its food. Although Wilkins was an advocate of agriculture in Australia, he cautioned that the necessary economic conditions must exist before agriculture could succeed.

At the same time as Wilkins was writing, there were others for whom the social function of agriculture was more important than any economic constraints it might face. The New South Wales colonial government established the Department of Agriculture in 1890 and, sixteen months later, the department reported on its operations. The report detailed the prizes being offered to farmers. These were judged not simply on yield or quality of produce, but on the 'cleanliness' of the land, and the general 'neatness and suitability of their house and farm buildings'. The department saw in this work the opportunity to foster a moral sensibility. Its goal was to 'raise Agriculture in New South Wales to the proud eminence as an honourable calling and an exact science which it has long enjoyed in the most highly civilised countries of the Old World'. The developing field of scientific agriculture could deliver a new class of technically educated, semi-professional workers and small landholders for the new century. It would be a mode of production more suited to a modern state than squatting or mining. Agriculture promised to return civilisation to the frontier.

Further shifts and expansions in the social purpose of agriculture occurred over the twentieth century. In the 1920s, agriculture became a reward for returned soldiers; towards the end of World War II, the hope that agriculture could increase Australia's population became vital to a government worried about invaders from the north; and in the 1960s, a productionist approach to agriculture was supposed to increase Australia's export income.

In *How a Continent Created a Nation* (UNSW Press, 2007), the historian Libby Robin examines the 'battler' ethos that emerged in this period. Robin shows how it was founded on a perception of the Australian environment as hostile and useless, and hence why the moral character of those who battled the land and made it grow European commodity plants was revered. The ethos was the basis of Australian national identity and culture for much of the twentieth century.

Values about the Australian environment are changing. Some of us are slowly abandoning the 'biological cringe', as Robin put it, and the continent is becoming unique and diverse. The environmental scientists David Lindenmayer and Mark Burgman, in *Practical Conservation Biology* (CSIRO Publishing, 2005), describe Australia as a continent in need of care, even in farm paddocks. This has undermined the moral basis for agriculture in Australia.

The battler myth holds less currency now, and many sections of Australian society – from environmental activists to farmers struggling with debt – have asked what the social benefit of agriculture is. In response, agribusiness lobbyists and rural politicians deploy the 'feed the world' slogan. What could be a more important moral imperative than feeding the world's poor?

Paul Myers, the former *Land* editor and now a freelance journalist, tried this line of argument in a column in the *Sydney Morning Herald* in April 2009. He suggested changing the word 'agriculture' to something else, because 'agriculture is no longer sexy'. In his opinion, this will help the world's undernourished, because agro-industrial farming could then get on with business as usual, without input from ecological scientists, economists, policymakers and the people farmers depend on, consumers of food. In short, Myers argues for retaining the current structure of commodity agriculture without dialogue or co-operation.

COMMODITY AGRICULTURE IS much more fraught than this implies. Marcel Mazoyer, a professor of agronomy at France's National Agricultural Institute and the author of *A History of World Agriculture* (Earthscan, 2006), writes that international agricultural commodity markets are 'residual markets glutted with surpluses that are often difficult to sell'. They are only a small share of world production, so they are not true global markets, but they have flow-on effects in local markets. Australia participates in a system of agricultural commodity trade between rich nations that disadvantages the poor and the hungry. Real food prices, despite the spike in 2008, are at historic lows – yet 850 million people suffer severe undernourishment, meaning they don't have enough food to cover basic energy requirements. Counter-intuitively, low food prices worsen the situation of the world's poorest people. This is because, according to the FAO, three-quarters of the world's undernourished are farmers or rural workers, and the lowering of world food prices through public subsidies for agriculture in rich nations pushes these farmers into extreme poverty.

Jean Ziegler, the UN's Special Rapporteur on the Right to Food, observes, 'You can go to the Dakar market [in Senegal] and find Spanish, French, German and Italian fruit and vegetables at half or one-third of the[ir] local prices.' Farmers in developing nations struggle to compete with highly subsidised commodities dumped on international markets. It becomes more difficult to raise enough money to maintain the family farm, to replace necessary equipment such as shovels and hoes. Often, desperate farmers try to extract more from their land and resources, and revert to low-capital methods of cultivation such as slash-and-burn, leading to environmental degradation. With deteriorating tools and declining resources, their surpluses diminish, and farmers are often forced to sell increasing portions of the food that they would normally keep for their family.

Eventually, these farmers lose their land and join the remaining third of the world's hungry: the urban poor. The current system of agricultural commodity trade creates a situation where low prices are detrimental to two-thirds of the world's undernourished people, while high prices are detrimental to the other third in urban slums. The Organisation for Economic Co-operation and Development (OECD) forecasts that if food prices spike again as they did in 2008, this will improve the income of farmers, agricultural workers and local economies as a whole in countries with large agricultural populations, such as India, Peru, Kenya and Sri Lanka. But countries such as Egypt, Nigeria and Haiti, with less agriculture and high populations of urban poor, will suffer. A spiral begins – of social and family breakdown, environmental degradation and landlessness, poverty and hunger. In this system, markets alone are ineffective in combating historical inequity and injustice.

WHAT THE LOBBYISTS never say directly is that the system of global agricultural commodity trade that they advocate, the system that disadvantages the world's poor and hungry, also does little to benefit the majority of Australian farmers. This is where Neville Simpson, at Bourke, comes in. He's not a typical cotton grower – his farm is not a subsidiary of a multinational company – but the shape of his story is familiar. It is not a particularly bad story, but it is common among Australia's family and small farms.

We were introduced to Neville by chance. A local pastor, interested in history, offered to take us to meet one of his parishioners. He said the parishioner was a long-time resident and farmer in the district, and would be a good source of stories and information about changes in the Darling environment. The pastor called ahead. Neville was attending a wake but would be home soon, and the pastor assured us that visiting wouldn't be a problem.

On the way out of town, we noticed a police car parked on the lawn in front of a weatherboard house, red and blue lights flashing, prompting the pastor to express concern about trouble in Bourke. On the highway, we drove past bleak rows of citrus tree stumps, cut in an effort to save them during the drought. There was scarred earth nearby, where others had been ripped out. Then we made our way into the distinctive corduroy fields of cotton country, their laser-levelled furrows converging in the distance.

Our initial awkwardness at arriving after a funeral was overcome when Neville's wife introduced herself, asked us in and offered homemade lemon cordial. At the back door, we stepped around a knee-high pile of lemons – part of a local farmer's surplus that hadn't made it to market. We took chairs on the front veranda, overlooking the black soil plain and a line of red gums marking the banks of the Darling. We all held tall glasses, sipping the sweet, cool drink. The hand Neville used to grip his was arthritic; the skin on his face was flecked with sunspots, red capillaries and other marks. His appearance spoke of years of hard work.

It wasn't long before he started telling stories about his life and about farming in dry country. He was born in a boundary rider's hut, and his parents worked a run in the red country between Bourke and the Queensland border. A 'starvation block', he called it. Wool prices declined steadily after the Korean War, and grazing in that dry scrub country was too much for Neville. He sold up with just enough money to move to town. When he had the opportunity to purchase land on the Darling in 1970, he jumped at the chance. It was a smaller block than he'd been on before, but it was rich black soil and close to the river. These characteristics, he thought, would 'drought-proof' his farming.

His idea was to grow feed for cattle, as beef was doing much better than wool. He planted lucerne but most of it didn't come up, and what did was no more than an inch high. He had no experience of cropping or irrigation. 'I made every mistake in the book,' he said, chuckling. Americans had recently bought land next to Neville, but they couldn't get a license to irrigate the cotton they wanted to grow. They saw Neville struggling, and said if they could lease some of his land and water they would teach him how to irrigate. That's when Neville was introduced to cotton.

In the 1890s, a government experiment farm had trialled cotton irrigation with artesian water at Pera Bore, twenty kilometres west of Bourke, but not in commercial quantities. No one else was irrigating the modern way in that region, treating cotton like an annual, and there was a lot of trial and error. They would benefit from a generous bounty that the government was paying to encourage farmers to go into cotton production, but they had to get a decent crop first. While they struggled to develop the cotton, Neville did contract farming for Edgell, the premium food company. With Bourke's climatic conditions being different to traditional horticultural areas, Edgell could 'keep the factory running all year round', providing supermarkets and consumers with the same commodity foods regardless of season.

Neville and his family planted potatoes, but the bagging machine designed for European soils couldn't distinguish between a potato and a clod of black soil-clay. It bagged more clods than potatoes. He tried tomatoes and grew a lush crop, but after they were packed, the western heat turned them to mush and they ran out the bottom of the crates. Neville's wife was heavily involved in their melon-growing venture, shipping 100,000 cases a year, but the market changed and labour costs were high, so it became unprofitable.

When cotton started paying in the 1990s, with the right varieties, technology and prices, it brought in a lot of cash. It became the saviour of Bourke. The British-owned Clyde Agriculture, the company that recently sold Toorale, built massive off-river water storages and irrigated tens of thousands of hectares of cotton. Irrigators joined the ranks of community leaders and, as the historian Heather Goodall notes, displaced many pastoralists. Neville and his neighbours expanded their cotton operations. Bourke was no longer running on welfare. That was until everyone upstream wanted to get in on cotton, too – that was, until this drought.

When I spoke to Neville, the drought had been going for five or six years, and he hadn't received any income for the past two. 'Now we're broke again,' he said, with his palms out. He managed a grim smile, but there was sadness in his voice. His family had had enough, but no one wanted to take on their property. Once conditions were good again, they'd look at selling.

Neville illustrated his story with a worn sepia photo of wool being carted on a massive dray at his father's run, an almanac of Australia with the years that Bourke was in drought carefully bookmarked, and a thin sheet of fax paper with the daily cotton prices, which he still monitored even though he didn't have a crop.

Later, I read in a collection of Indigenous oral history from the Bourke district that Neville had earned respect among the labourers and cotton chippers who worked on his property. One worker spoke about Neville giving his precious household rainwater to the chippers, commenting that he was the 'only cotton grower' who did that. Perhaps Neville wasn't as distant or ruthless as corporate cotton growers; perhaps farming isn't just about money.

We need to be honest about the role agriculture plays in Australia and start developing support for a fairer system for Australian farmers, the environment and farmers in developing nations. Trying to force a productionist culture of farming isn't benefiting many people.

Neville Simpson's story maps the fortunes of Bourke. Cotton ended up like all the other commodity industries that promised wealth and ended in ruin: first, squatting and wool; then the meatworks, which took over for a while, but whose factory building stands derelict at the edge of town. Locals give different reasons for cotton's demise. Some agree with Senator Bill Heffernan when he says the water stops at the Queensland border now; others say it was never going to last long when dependent on a river as variable as the Darling. Neville doesn't blame Queensland, or any other upstream irrigators lured by the promise of cotton. Instead, he says, 'When it rains, the river flows.'

As we were preparing to leave the property, Neville, his wife and the pastor stood under the carport and spoke gravely. 'Looks like we had a suicide in town today,' one said. That's what the police had been attending when we saw them earlier. The conversation moved on to the hardship of mental illness, and the need to ensure the community received the support it requires.

The owner of the motel I stayed at on a subsequent trip to Bourke mentioned she was sad about another suicide. The black dog stalks Bourke.

AUSTRALIAN FARMERS ARE among the least-subsidised farmers in the OECD, but they compete against highly subsidised commodities on international markets. They benefit from access to machinery and research, but the inputs are high and the profits marginal. Australian Bureau of Statistics figures for 2006/07 show nearly one-quarter of Australian farms had an annual production value of less than $22,500, and nearly half brought in less than the average male full-time wage of $57,000. Some farms do well. Around two-thirds of cotton farms had agricultural operations worth more than $500,000 for the year; but, as Neville's experience shows, not even cotton is reliable, and water is over-allocated in our stressed, life-giving river systems.

In a report commissioned by the FAO, Marcel Mazoyer found that only enterprises with the capital to continually find new competitive advantages can survive in a market where prices are artificially driven down. Often this means 'externalising' costs to the environment. Frank Vanclay, a professor of rural sociology at the Tasmanian Institute of Agricultural Research, says Australian farmers either make do with less, 'flog the land' or sell.

He argues that the emphasis on restructuring the rural sector in line with 'economic rationalism', and neglecting support for the many values and functions of Australian farming, results in social hardship and environmental degradation.

The ways of tackling world hunger can appear hopelessly contradictory. If food prices rise too quickly, millions more of the urban poor plunge into a position of food insecurity and near starvation, yet subsidies that drive down the prices of a few market commodities in wealthy nations destroy the livelihoods of two-thirds of the world's starving people. If public subsidies are withdrawn from the farming sector in some wealthy nations but not others, the small and medium farmers who lose subsidies end up in relative poverty. Providing public subsidies can keep 'bad' farmers in businesses and contribute to environmental degradation, but taking away subsidies can lead to extra pressure to extract more from the land, and cause environmental degradation. All such scenarios of interrelated failure rely on the market alone to solve political, historical and cultural problems.

Scarcity of food is not the reason that millions of people around the world suffer. There is enough food, but the current structure of global food production is inequitable and strongly guarded by the few who benefit. In July 2008, amid the price spike in international commodity markets, the world's third-largest pesticide and seed company, Syngenta, released a media statement, 'Hungry mouths to feed: The role of Australian farmers in finding a global food solution'. It played on empathy for the world's poor and hungry to sell the company's products to farmers who only contribute a fraction of global food production. It pointed to food riots in developing countries and tried to elicit the perception that wealthy nations need to produce more food to solve the problem of world hunger. It played on the goodwill of Australian farmers, and urged them to buy chemicals, with statements such as, 'Australian farmers will need to better their existing rate of productivity improvement (and importantly yield improvement) in order to stay a step ahead of growing demand.' Journalists could email Syngenta to obtain a photo of Syngenta's general manager, Paul Luxton, with the caption 'Australian farmers who embrace new technology can help ease the current world food crisis.'

The Swiss professor of sociology Jean Ziegler, in a report for the UN, cited FAO data that shows the world already produces enough food for its current population, and 'easily' up to twelve billion people. Alternative modes of producing and distributing food that move toward a more equitable system, such as fair trade and food sovereignty, can be hard to imagine as global alternatives, because we haven't seen them functioning on that scale yet. We can end up feeling overwhelmed by the weight of historical injustice, the power of vested interests, and our flawed relationships with non-human nature.

However, the American farmers Sharon Astyk and Aaron Newton, in *A Nation of Farmers* (New Society, 2009), argue that wealthy nations have an obligation to begin changing the current structure of agricultural production and the global commodity trade. They argue: 'Beginning from the assumption that greater equity is impossible naturalises disaster — it says that the reason people starve is because we can't do anything about it, and it makes it easy for us to wash our hands of the whole project of justice.'

While global poverty and hunger can seem too complex and so large-scale that we feel disempowered, we do know about Australian farming; and we know that changing the current system of agricultural production in wealthy nations is a first step towards allowing farmers in developing nations to feed themselves, and a first step towards making a meaningful contribution to the injustice of world hunger. In Australia, we can start to think about a system of 'regenerative agriculture'. George Main, in *Heartland* (UNSW Press, 2005), suggests we undertake this because it 'acknowledges a painful history of suppression, fragmentation and disorder. Connectivity is acknowledged and nurtured.'

LANDHOLDERS AT THE Macquarie Marshes have already started to turn their backs on productionist agriculture. In August 2008, a month before the federal and NSW governments purchased Toorale Station, these governments bought part of Pillicawarrina at the Macquarie Marshes, another historic property. There were no rallies in the nearby towns of Warren, Quambone or Carinda. Most landholders reacted positively to the purchase: they know the Marshes can't survive without water. They know the environmental degradation and social hardship associated with the current system of industrial agriculture threatens the long-term viability of agriculture itself.

The state needs to play a major role. It is not fair to outsource environmental problems to individual farmers. Most of the Macquarie Marshes are private land,

and this necessitates dialogue between reserve managers and private landholders in the complex, interconnected ecology of the floodplain wetland. They are moving beyond a simplistic opposition between production and protection.

At Toorale, National Parks have assigned four rangers to oversee the management of the land. That means up to four families will be living there; when Clyde Agriculture owned it, there was only one resident family. The National Parks staff at Bourke are generous and capable. Many are traditional owners in the western area, and this adds another dimension of expertise and experience.

In 2008, I travelled to Bourke with the anthropologist and ethicist Deborah Bird Rose, and we spent three days with Phil Sullivan, a traditional owner and National Parks officer. Phil is working on a research project to collect and publicise Indigenous values for water. He is ensuring that Indigenous people have a say in the management of rivers, because 'water is life'. For too long, he says, non-Indigenous culture has separated the human world from nature. He is working to restore relationships of care and mutual benefit.

Similarly, in an article in *The Economist*, the head of an Indigenous organisation in Bourke said he is excited about the opportunity to get back on country at Toorale, to work with the National Parks officers identifying and preserving sites. He added, 'I hope that one day a community like Bourke will be run by Aboriginal people.' Hope is returning to Bourke.

Regenerative agriculture, the dialogue between state reserve managers and private landholders in the Macquarie Marshes, the culture-changing work at Bourke: all are a different way of thinking about the Australian environment. Deborah Bird Rose, drawing on Levinas, uses the phrase 'nourishing terrain' to describe Indigenous relationships with, and notions of, the environment. She has written, 'Country is a living entity with a yesterday, today and tomorrow, with a consciousness, and a will toward life. Because of this richness, country is home, and peace; nourishment for body, mind, and spirit; heart's ease.'

That is a much more appealing position than a system which results in ecological degradation, stress and fear, poverty and hunger. Since Rose wrote this, in a report for the Australian Heritage Commission in 1996, the concept of nourishing terrains has been widely quoted, from land managers to historians and geographers. It might become the way Australians understand their environment. The function of agriculture would have to conform to this way of seeing, living and working. Perhaps the Australian experience could even be something that other wealthy countries exporting agricultural commodities could learn from.

IN FEBRUARY LAST year, Bourke received two hundred millimetres of rain in one day, two-thirds of its annual average. The town flooded, not because of the Darling River, but because of the sheer volume. The footage on TV, two months later, showed the country around Bourke in the best condition I'd seen it for years – a good time, it seemed, for Neville Simpson to sell up and move on.

I called Neville recently to see if he had made a decision. He sounded weary over the phone. Just before last year's rain the bank had forced his family off the farm. His water license had been permanently reduced without compensation, and the bank withdrew its backing. At eighty, he walked away with no assets. He had put every cent he made back into the property. 'It was bad management on my part: I should have bought a unit on the Gold Coast,' he said wryly. Before the drought the farm was worth around ten million dollars; now he is living on the pension. Fortunately a family friend put him and his wife up on their property thirty kilometres outside Bourke.

Neville's son, who had also lived on the farm and had shared the debt, now has a permanent job with the shire council. He has taken out a loan to build a small house in Bourke. Neville said he has found security in old age. The family will live together again when the house is finished.

References available at www.griffithreview.com

Cameron Muir is an environmental historian undertaking doctoral studies on regional New South Wales. He is based at the Fenner School of Environment and Society at the Australian National University. Cameron has had work published in *Australian Humanities Review*, *Redoubt* and *Voiceworks*, and his story 'Is your history my history?' was published in *Griffith REVIEW 13: The Next Big Thing*.

A taste of home

Learning to cook

Rebecca Huntley

THE first time I visited the Matthew Talbot Hostel, in the inner-Sydney suburb of Woolloomooloo, was Christmas 2005. My husband had received a thirteen-kilo ham from his boss. As I'd already ordered a five-kilo ham, we thought we'd give the larger chunk to the hostel. Walking down the alleyways towards the entrance, I recall tensing up as I spied downcast men sitting alone or in groups on the footpaths. I forgot for a moment where I was going and what I was doing, and felt like turning back. But I didn't, and dropped the ham off to a worker sitting behind a desk protected by glass. He said it would be much appreciated by the guys. I wondered how the residents felt about eating off-cuts from the tables of the over-catered.

Four years after the ham drop, I returned to the same spot, albeit to the Ozanam Learning Centre, a sleek $20-million institution run by the St Vincent de Paul Society, and connected to the hostel by a walkway. Food was bringing me back, but this time I was there to observe one of the centre's cooking courses for homeless people.

My taxi pulled up just past the door, at a smallish courtyard where groups of men and some women sat, surrounded by bags of possessions and the odd shopping trolley. 'This is where you want me to drop you?' the driver asked.

I presented at the front desk, a nicer version of the hostel's reception, with the same security glass in place. Jamie Lehn, Ozanam's activity officer and my host for the day, happened to be arriving as I asked for him. I had chosen a busy time to visit; they were launching Missing Persons' Week that morning.

We headed up in the lift. The cooking class I was there to observe was due to start in half an hour. 'I think you'll be surprised,' Jamie told me.

IN 2007, BLACK Inc. published my book on food and inequality in Australia, *Eating Between the Lines*. Its central premise is that the way we eat, our relationship to food and cooking, reflects social trends, particularly social inequality. One of the omissions from the book was an analysis of the diet of homeless people. As part of my general reading on food and inequality, I came across the work of Sue Booth, who has done some of the more in-depth research on Australia's homeless people and food. In 'Eating rough: food sources and acquisition practices of homeless young people in Adelaide' (*Public Health Nutrition*, 2006), Booth describes the food intake and sources of 150 homeless youths. She found that eating three meals a day was unusual: most said that they ate only once a day. When they did eat, it was a limited diet, lacking in foods from the five food groups, especially fruit and vegetables. Almost three-quarters of the young people Booth spoke to reported going to sleep hungry once a week or more. Stolen food was as popular as food distributed by welfare agencies or acquired through begging. Booth found some of her interviewees used jail or overnight detox facilities as a source of relatively high quality food: the chance to eat stews, lasagne, and fish and chips under a roof, at a dinner table.

In November 2008, Adele Horin wrote for the *Sydney Morning Herald* about the opening of the Ozanam Learning Centre, with much fanfare, by Prime Minister Kevin Rudd. A photo of Alan, a star pupil in the centre's cooking course, accompanied the story. Horin described how important it was for the long-term homeless to learn basic living skills such as cooking before moving into a flat or group home. For someone like Alan, cooking could also help rebuild self-esteem and confidence after long periods of struggling to survive. Here was a stark example of what I was trying to explore in my book: how food and eating reflect social disadvantage, but also how engagement with food can enhance social connections and social capital. After reading Horin's article, I was determined to understand more about Ozanam's course and the impact it was having.

BEFORE JAMIE COULD give me the tour of the centre, I had to sign in and get a personal alarm. If something untoward happened, I would trigger the alarm and an unbelievably large security guard would come to my aid. Jamie assured me that it was simply a precaution. 'Ninety-nine per cent of the people here are good people. The other one per cent, they have problems, but you can understand why.'

Behind the reception desk I could see an open recreation area with rows of computers, a ping-pong table, and coffee- and tea-making facilities. Jamie gave me the tour of an extensive art and craft room, a lovely little library and rooms for one-on-one tuition, therapy classes, and the centre's movie and book clubs. In one of these rooms, a man practiced guitar. Upstairs there were recording studios and various instruments.

The centre provides services for both the housed and those living rough. They were almost exclusively men the day I visited, although Jamie told me the centre was doing all it could to get women involved, including starting a special women's group.

As I entered the computer lab, a tall and imposing Islander with a backpack and heavy boots spied me. He did a double-take and greeted me with a perky hello. He returned later to give me some sage advice, whispered in my ear: 'Don't take a poor man home with you' – advice my mother gave me on more than one occasion.

Finally, Jamie showed me the kitchen, the venue for the cooking classes, where at the centre's opening the Prime Minister had been photographed stirring pasta with the students. It was small but well equipped, light and clean, with a window onto the industrial backstreets of Woolloomooloo; a simple table with chairs at one end of the room, a sink for hand washing, two ovens, two stovetops and white melamine benches.

The class was full, as they always are. It took a while to kick off, with Jamie having to round up those who had already signed on to participate. People began to gather: mostly men, but there were two young women as well, a tall brunette and an older redhead. As we began washing our hands, one of the centre regulars, 'Dave', started to cross-examine me. Why was I here? Was I coming back?

I told him I was writing a story about the class. He replied that maybe I should get involved with the women's group, help run the book club or start a writing group. I guess he'd seen lots of people like me come, stickybeak and go, and was determined to recruit at least one of us into an ongoing relationship with the centre.

The dish of the day was vegetarian lasagne, a recipe from Victoria Hansen, the course creator and co-ordinator. Jamie pulled out the ingredients and the pots and pans as the students surveyed the recipe. And then, quietly and seamlessly, people got to work. The redhead started off the béchamel sauce and the brunette assisted. An older man, with Elvis-like slicked hair and cowboy boots, chopped the zucchini and garlic. Another guy, tall and quiet, began the tomato sauce. I introduced myself to 'Sam', who pounded some salt and pepper in a mortar and pestle. He told me he liked to cook but often forgets the ingredients for recipes and hence what he needed to buy to make them.

As the students worked, I stood to the side, determined not to ask too many questions, just to help out and be part of proceedings. The conversation focused on everyday challenges: the politics of group homes, Sydney's high rental costs, the police. Each student's difficult history, recent and long-gone, sometimes edged into the conversation, then faded away. Like the brunette, who told me she was eighteen (she still had braces on her teeth): we started talking about how to cook spaghetti. I told her I was Italian; she said she was as well, but on her father's side. 'After he died, that's when I left home.' I found out the guy making the tomato sauce was her partner, 'Ian'; he was thirty. I went over to talk to him, to ask him what he enjoys about the cause. His response was blunt: 'I prefer to make this than eat over at the hostel.' I asked him if his mum was a good cook. 'Sure. She was. I haven't seen her since I was seventeen.'

As the students started layering the vegies, sauces and pasta sheets, I talked to 'Kim' and 'Susan', older women who arrived at the class late. We talked about the challenges of cooking for one, their frustration that shops don't offer small sizes of things, especially vegetables such as pumpkin. They worried about wasting food and trying recipes that didn't work. Kim believed that if she learned some cooking skills she'd be able to eat well on less money. Susan was sceptical. 'I went to Hungry Jack's last night and had a whole meal for five dollars. I couldn't cook that meal myself for that amount.' I asked them if they ever watched cooking shows. Susan liked *Huey's Cooking Adventures*, but couldn't attempt the recipes. 'He cooked a dinner for two with salmon the other day. He used twenty dollars' worth of salmon for one dinner. I can't afford that.'

Once the lasagne was in the oven, the class dispersed and a few of us were left with the washing up. Sam struggled with the flick mixer. 'There's no hot water left.' I flicked it to the left and he started to dry the dishes as I scrubbed.

The prep done, the dish baking, Jamie and I sat down to talk. Why are the classes, held each week for eight weeks, so popular? Along with cleaning, budgeting and other household skills, housing applications require skill tests in cooking. Learning to cook helps homeless people 'move up the ladder', as Jamie put it. But more than that, it gives them a 'taste of home'. I had seen that. I wondered how many home-cooked meals Ian and his girlfriend had enjoyed in the years since they left their families.

I said goodbye to Jamie, promised Dave I would return and left my personal alarm at the centre's reception. I was starving, and walked a few blocks towards the trendy East Sydney restaurant area on the other side of William Street. I spied a busy noodle bar. Outside, well-heeled office workers were lining up to pay fifteen dollars for salmon and chilli stir-fry and rice-paper rolls. I forked out almost twenty dollars for my lunch.

A FEW WEEKS after visiting Ozanam, I talked to the chef and teacher Victoria Hansen, who designed the course and runs its introductory class. She became involved when Jamie contacted her after seeing her no-nonsense publication *First Principles: The Basic Cooking Handbook* (VLH Enterprises, 2003). Hansen jumped at the chance to teach people with few skills and few resources a love of cooking. 'I felt there was so much out there being taught by celebrity chefs that wasn't touching on the funda-mentals of cooking, and there are a lot of people out there who don't even know how to boil an egg. People have lost the art of how to prepare food.' Hansen's approach to the course is to teach the fundamentals – meals that are simple, inexpensive and nutritious – and then suggest variations. The biggest barriers to cooking, she believes, are fear of failure and the expense of cooking something new and different. She advocates foolproof recipes, so students won't waste time and money.

At Ozanam few, if any, pictures of food are shown to the students. 'You first teach how to master the preparation, so they can see that it tastes good. The next stage is how it looks on the plate. People who don't know how to cook don't want a picture. If it doesn't turn out like the picture, it's discouraging.' The aim is to encourage people, to teach them to fend for themselves. 'Making stuff from scratch gives you courage,' Hansen tells me.

I asked her what she thought the students get out of the course. 'They want to make a go of their lives, do something different. A lot of them are there too because they just want a meal, which I totally understand. Once they have tasted – the look, the shock, that something so simple can taste so good. It's fulfilling for me to see them transformed. They do go away and give it a try.'

More than that, Hansen believes learning to cook gives students a greater sense of connection to others. 'It's taking your life into your hands when you start to prepare food, because it is the thing that sustains us. The meal is the one place all human beings come together.'

AT THE BEGINNING of my tour of the Ozanam Learning Centre, Jamie had told me I would 'be surprised'. I was. I expected that the issues homeless people face in learning how to cook would be different to those that the rest of us face over food and cooking. But they are the same, even if the barriers are greater for people like Dave and Ian and Kim.

Without resources, skills and incentives, cooking seems a waste of time and money, something that other people do. A five-dollar dinner at Hungry Jack's seems a better option than cooking our own lasagne. And yet, when we master the basics, food and cooking can provide us with more than a meal: greater fulfilment, the chance to move up the ladder and, for some, a taste of home.

Rebecca Huntley is a social researcher; the director of the Ipsos Mackay Report, Australia's longest-running report on social trends; and the author of *Eating Between the Lines: Food and Equality in Australia*.

Hospitality

A cautionary tale

Jim Hearn

I was fifteen when I stood inside my first restaurant kitchen. My mother, who had replaced the red dirt of Mount Isa for the red lights of a different town, had organised this, my first full-time job. I had no comprehension of what was required or of the implications for my future there.

Oliver's Seafood Restaurant makes no dining lists of substance; it has no stars, hats or write-ups. Like most restaurants of its kind, Oliver's has closed and re-opened half a dozen times under different names and with different culinary dreams.

Dreams drive hospitality. While some people like to think of it as a component of the service industry whose responsibility it is to address the needs of the body, for those on the inside it is a weird and sometimes wonderful dreamscape of ungodly hours, ridiculous pressures, unkind owners, absurd customers, torture, humiliation and occasional moments of brilliance. The obscure thrill of putting it all together on the night, of getting all the sections of a busy kitchen firing, is like no other I know.

Glenn, my first head chef, was fresh out of cooking school and had an unabashed passion for red cordial laced with methylated spirits. On the rocks. He had a ginger handlebar moustache that never quite worked and a temperament I would come to understand as a particular type. He was a busy, nervy bundle of energy and somehow, despite his jitteriness, easy to get on with. Glenn figured that if life had sent you inside his kitchen, you were pretty much fucked and there was no reason to make things worse. He possessed a capacity for kindness, a capacity that evaporated every lunch and dinner. The weight of being vaguely kind outside service hours demanded his inner anger be given free reign as the orders rolled in. Hell hath no fury like a stressed-out, angry head chef in a busy, no-good restaurant. Forget fine-dining celebrity head chefs – in the out-of-the-way suburban business-lunch trattoria or city-fringe motel diner, hell has a name: Chef.

AS I STOOD in front of my first open cool room, I couldn't move. My senses were assaulted by an unreasonable number of smells. It's an

PREVIOUS PAGE: *There's no time for funny business (detail)*. Photographer: Jeremy Nance from the *Sydney Morning Herald* Shoot the Chef! 2009 photographic competition.

odour which recurred over the next twenty-five years in inexplicable ways: as if this cool room were really all possible cool rooms, its stored ingredients – its sauces, produce, pastes, meats, condiments, cheeses, seafood and mould – melded into one archetypal olfactory sensation. I've since cooked in Middle-Eastern and Asian restaurants, and while the food tastes and smells different, the cool rooms all smell like Oliver's Seafood Restaurant.

'Shut the cool room, faggot,' said Glenn in his semi-kind, before-service voice.

I slid the door shut and turned back to the kitchen. Glenn, wearing a gee-whiz-kiddo stare, asked, 'Where's the butter, fag?'

This thing about being a faggot: it's offensive, a puerile and ridiculous turn of phrase, but in Glenn's world everyone and everything was a faggot.

'Sorry, chef. I forgot.'

'Jesus fucking Christ, fag,' Chef Glenn mumbled as he barged past me into the cool room, ripped the butter from the shelf and slammed the door so hard the bell rang. The bell on the outside of every cool room door is there to attract attention if someone gets locked inside, which happens a lot more than it should.

'But-ter,' Glenn droned, as though he had a speech impediment, as if I didn't know what butter was.

Standing there in my first restaurant kitchen, suffering beneath the gaze of my first kitchen humiliation, a pubescent fifteen-year-old kid who had just left his eighth worse-than-average school having failed half his subjects, I felt like a fool. I was a kid with parents in the throes of a divorce, with no idea of what I was doing or how I should be feeling, what consti-tuted being an outright moron or apprentice of the year. I was not alone. Many people like me find employment, and a life, inside their version of Oliver's.

'You're a moron,' said Chef Glenn, testing me.

'Yes, chef,' I mumbled.

'What?'

'Yes, chef,' I yelled vigorously.

Chef Glenn stared at me, nodding, semi-impressed.

I couldn't help but smile as the waiter ripped the first couple of lunch orders from his order book.

BEYOND THE COOL room, the second archetypal smell of the kitchen is that of a just-blown-out bamboo skewer. Chefs all over the world light skewers off pilot lights or a gas burner to light the burner they need at that moment. Then they shake off the flame, or blow it out, or snub it in the wok station water, and the smell...I don't know what it is; I just know it's the same everywhere and means the day has started and it's time to get to work. Everyone knows the six boxes of matches and two lighters that were neatly placed in their service-is-coming-get-ready-positions are out the back in the storeroom with the empty beer bottles, wine glasses and ashtrays. They're shoved under dirty aprons or empty cardboard boxes and they'll be retrieved in time, by the apprentice, before service and well before the owner or restaurant manager turns up. It's not that every service ends in a celebration; it's just that at the end of every service, two things happen: chefs get changed out of their kitchen whites and, in the process of this changing, this metamorphosis back into street attire, the service that has just gone unravels in their minds and in their conversation. And if it has been a massive service, if you've just been slammed, you'll be having a drink and chatting, laughing, taking the piss and generally talking it up, and this moment of bonhomie has been known to extend beyond the mechanics of getting changed. There's a tipping point, perhaps as simple as agreeing to the second drink, that signals to everyone: get the fuck out now if you have to...and if you don't, you'll be going to bed with the words *Where the fuck are the matches, faggot?* ringing in your ears.

My first pay cheque stretched far enough to buy a carton of beer and some fancy Italian deli food which I thought I knew about. Of course I got outrageously drunk and danced and vomited and drank some more – on a beach in Townsville, home of Oliver's. I have a vague yet pressing sense the night ended in tears: childish, transitional tears about the confusion and hopelessness of my past life which necessarily involved my parents and siblings. Alcohol seemed to bring me undone in this fashion each time I drank. I remember older, wiser heads patting me on the back, ruffling my hair, telling me that it'd all work out: that I'd be fine; I'd see.

Yet what I came to see was that despite working like an adult, taking on responsibilities and more hours than was reasonable or sane, I was a kid who couldn't come to terms with what had gone before. I was both a child and a teenager, and my family crisis informed who I was, and was probably the reason I drank so much so young. Hospitality for me is and has always been a transitional space: kitchens come and go; people slide in and out of a chef's life 'until the next gig'. My capacity to leave, to not stick it out, to come undone and move on, was a fitting continuation of a pattern from childhood.

WHEN MY PARENTS divorced, it was my mother who left my father. This fact, about who leaves, is perhaps the most telling signifier at the end of every relationship. Many couples like to say that it was mutual, that each of the parties came out of it with their pride and self-respect intact, but anyone old enough to have fallen in love knows that's bullshit. Someone gets done over and someone does the doing over. So, as the song goes, my mother, after having six children and not missing a Sunday mass in sixteen years, well, we were told that she'd been bold with Harry, Mark and John…and before the song closed out, she found herself working in brothels around Kings Cross.

The thing about being a prostitute is that for most people, it's a fantasy life, something only ever imagined, read about or seen on television. But, backstage at least, it's like most other jobs. There are the usual dramas, boring bits, good days and bad; but there aren't a whole lot of people you can talk to about what happened at work. Unless you're my mother and you don't care what other people think and you never pause to reflect on how your actions might affect other members of your family. No, if you're my mother, you tell anyone who'll listen about what Bob did to Jane and she did to him and they did to her. My mother had no capacity to wonder about how her actions might affect others; this fault line in her personality ran deep, and grew wider and longer, as if she hoped one day she might wake up and find she'd split in two. Only she never did, and it was left to me to work out what it meant to have a mother who worked as a prostitute.

As such, my early years in the kitchen were a kind of double life. It wasn't that I felt that I needed some mystery, some other secret world to feel good about myself or keep people guessing – I wanted nothing more than to be able to focus, draw my attention to what was happening in front of me – it's just that this other world, the world of my unconscious, my childhood dreams and memories, seemed to wash over everything, creating a haze through which I would attempt to pull myself together in order to survive financially and, with any luck, learn something about what it meant to be a chef.

CHEF GLENN MOVED on from Oliver's, gaining employment as a senior *chef de partie* at a resort in Cairns. He was very happy about this; the idea of joining a large brigade seemed to bring his aggression levels down to a level more commensurate with his talent. He could hide out in a big resort kitchen. There was a two-week respite from his service temper-tantrums before Chef Ivan arrived.

Chef Ivan was a good chef – unlike Chef Glenn, who had absolutely no idea what he was doing. Chef Ivan had sailed from Tasmania for the tropics, and hit the kitchen with gusto. He threw all of Chef Glenn's preparations in the bin and started again, working overnight to re-stock the cool room with the *mise en place* of a new and much better menu.

Chef Ivan actually cooked. He combined ingredients in ways beyond Steak Dianne and Vegetables, and when I arrived at work early one morning and opened the cool room I was amazed at the sight of buckets of yellow, green and red food: broths and soups, curries and sauces. And although the cool room still smelled the same, the food tasted better. I stuck my fingers into everything, pulling out globs of yellow, coconutty seafood, tomato soup, green curry.

Things were on the up at Oliver's. I was living with a large, gracious Maori family not far from work and I was looking forward to learning something about cooking. Then my mother rang.

Since leaving me and her two youngest children, 'She' – as the family called her – had gone to Sydney and hooked up with an ambitious young café owner. Apparently there was an opportunity there for a first-year apprentice chef and, out of the blue, She thought of me. And the thing about being an apprentice chef, which I didn't understand then, is that any employer will give you a start if you show enthusiasm – apprentices are basically slave labour.

When She rang I was still a hopeful teenager who wanted life to work out. I could still think the best of people and remembered my mother as a mother and not a sexualised, super-thin person with desires and ambitions. I was pleased she rang, proud that she had thought of me, that she wanted her son to help out. And I was more eager than I should have been to show her how productive I could be.

AS I STOOD in front of the second commercial cool room of my apprenticeship, the familiarity of the smell imbued me with confidence. *I know this game*, I said to myself. *I can do this.* And as I dived into the cool air, I was a little surprised, a little underwhelmed with the produce that lined the shelves. There was a sense of the room being disorganised and basic, more a large refrigerator used to store bulk produce than a repository for a restaurant kitchen's preparations. And that was because The Pantry was more a deli-cum-café than restaurant, more a place where people eat everyday than a restaurant with table service and a three-course menu. The experience of working there – which again turned out to be for a short time – left me with a conviction about what I didn't want in my life. I didn't want to work in a business-lunch, tourist-friendly take-away joint (however good the Reuben sandwiches were), and I never again wanted to work with my mother in the same kitchen.

My mother was becoming something of a disappointment to me. In no time flat she married the owner and the whole new-head-chef, new-dad thing wore wafer thin. So I found myself stopping off at the pub on my way to work, on the way home from work and during the day for a quick fix. I was on a pack and a half of Winfield Reds a day and I missed the crew from Oliver's. But I never thought of going back. There have only been a couple of kitchens out of a couple of dozen where I've gone back after a sustained break.

What The Pantry taught me was how to burn bridges rather than piss around and pretend everything was going to work out in the end. Quitting a job by becoming a reckless alcoholic or addict became a routine I would repeat many times when the pressure, or the crew, or the owners, or the girls – and the girls were the worst of it – got too much to bear.

Step One: Turn up to work half an hour late and smelling of beer three or four days in a row. It doesn't matter how highly the owners thought of you, they've already done the sums on what it's going to cost to show you the door and generally they have the cash ready in their pockets.

Step Two: On day four or five, come in an hour and a half late and you'll find them shaking their heads in disappointment. Of course they have to let you go and of course you're disappointed, surprised, you thought things were going really well – you were just getting into a routine. And this news of a routine where you come in late and drunk every morning has them reaching into their pockets, which is always the trick, pushing the scene until they show you the money rather than just talk about it.

Step Three: Go to the pub and get laid with the only waitress you haven't bedded yet. In the morning, after you've said your goodbyes to Betty, tally up the mental list of joints you know are advertising for a chef or apprentice or cook and then, when you've narrowed the list down to two shops, go to the pub. It's going to be a new pub, a whole new scene closer to the restaurant you think you want to work at. Here you can scope out the new joint, watch the suits roll in for lunch and figure out how busy the place is.

THE OBSERVANT READER will note that this is not a recipe for success, a guide to rising through the ranks of a fine-dining scene: these steps are survival techniques for the early years of hospitality, a means to paying the rent, keeping beer on the table and ensuring a regular supply of sexual partners who share, a least for a few hours a day, the joys of such an existence. And it is only ever a plan for the stage of life known as the pre-serious-relationship phase. Not all chefs understand this, and it's a shame: their partners or pets deserve better. The serious-relationship partners are generally good people, often at a critical child-bearing age, and it's the chef, who has become accustomed to life as a barfly and lowbrow kitchen scum, who doesn't realise that such a life is only suitable for the single person. It is not a road to nowhere: young drunken chefs do after all mature into older drunken chefs.

It wasn't long after I left The Pantry that She took the café owner for a good portion of his net worth – and, frankly, he deserved everything he got. Some proprietors aren't worthy of the money they make. By parking themselves in a busy tourist spot where every day new and unsuspecting customers stumble through the door, exchange rates buzzing too slowly through their weary brains, the proprietors are able to cash in. You don't get many repeat customers.

The next time I saw my mother, we caught up on family gossip over a Medium Big Mac Meal in the bar of a brothel in Bondi Junction.

THE BONDI HOTEL is a landmark beachfront building in Sydney's east. It seemed more of an institution twenty years ago than it does today, somehow less try-hard and more gracious. When I started work there, it was like walking into a dreamscape. The place didn't just smell of beer and cigarettes: the odours were ghosts which held the joint together. It had nooks and crannies, rickety staircases that led nowhere, bathrooms which weren't used anymore, at least not for what they were designed for, and a drug-dealing scene that was the envy of all the other pubs in the area. We had a twenty-four-hour licence and, come public holidays and long weekends, the hotel never closed. Christmas was a blast. This was before pokies ruined everything great about public bars. People played pool and talked, drank beer and took the piss. They also fought and argued, ate and fell asleep.

In the late 1980s, the hotel was a haven for Kiwis fresh from New Zealand, Pacific Islanders and leather-clad bikers. The three main bars were run by these sometimes warring, but generally peaceful, factions. The Kiwis had the public bar, the Pacific Islanders had the poolroom and back bar, and bikers had the front bar. They controlled the sale of drugs and were responsible for their own security. Staff turnover was high. Not everyone was into three-in-the-morning knife fights, glassed-in-faces and syringe-clogged urinals. For me, it was heaven.

The kitchen was busy and the food was basic. This was not fine dining, nor a weight-watchers' club; there were some very large people about who had a penchant for slabs of beef with buckets of vegetables and sauce. Food costs were out the window. This was about keeping the blood-sugar levels of very imposing men at a level whereby they were friendly to me. I was a kid from Queensland who was happy to maintain the status quo. They let me know what they liked and how they liked it, and I did my best to provide. It was a simple arrangement, and I'm pleased to say I got to call some of these people friends. I moved on to another kitchen, of course.

THOSE SLABS OF beef at the Bondi Hotel bring me to a recipe, or more accurately a method of cooking.

Steak

Ingredients:

Any tender cut of beef.

Method:

Let's be clear about this: steak is beef. It is not pork or lamb or chicken or duck or fish. Steak comes from beef cows and the best for eating are Angus. Far too many chefs and backyard cooks think they know far too much about beef and, quite simply, they don't. Cooking beef is one of the most underrated skills a chef can acquire. How hard can it be? Let me tell you, cooking beefsteaks – particularly the thick, fat eye fillet or tenderloin – how the customer orders it is an underappreciated skill. Every steak is different. There is no such thing as a perfectly timed steak, and I know plenty of restaurants buy portion-controlled tender cuts to try to get around the difficulty of inconsistent amounts of meat, but trust me, these aren't worth the premium.

Cooking steak is a feel thing; you've got to develop a sense of touch around the flesh of animals that becomes fine-tuned, accustomed to degrees of cooking. A good chef knows when a fillet steak is medium-well or medium-rare or just on medium, and they know because they have developed a sense of touch. By squeezing a steak while it cooks, a chef can assess the stage it's at. There's no quick and easy way to get to the point of consistently 'knowing'; you just have to develop the skill.

One of the best ways of teaching a chef how to cook a steak is to put it on the menu as a sliced dish. For instance, the dish might read: Sliced Angus

Tenderloin with Yam Puree and Shitake Mushroom Jus. Now, the reason the tenderloin is sliced is because the chef, which in this case is me, is training an apprentice to cook steak. I want Cam, or Marty or Rani or whoever it is in the next joint, to gain confidence in their ability to cook steaks by seeing and analysing the inside of every one they prepare. So, after they've seasoned the lump of beef with salt flakes and white pepper, and seared all the surfaces of the steak in a cast-iron pan, the fillet goes onto a clean tray and into the oven.

Here's the other thing: every oven is different. It doesn't matter if Chef Pete puts it in for two minutes and Chef Jane says three-and-a-half and Sous-Chef Donny goes six; every steak and every oven is different, and the only way to get a steak medium-rare, as the customer ordered it, is to be able to pinch the steak between your thumb and forefinger and know whether it needs a while longer in the oven or it's done. And done means it's ready to be rested. The steak needs to be transferred to your protein tray in a place where it is going to stop, or very quickly slow, cooking. It needs to be rested so that the blood will congeal or drain from the piece of meat before it gets plated.

Before I let the apprentice slice the fillet, generally into three thick pieces at a nice angle, we look at the steak, we touch the steak, we talk about the steak and watch the blood soak into the tea towel sitting on the protein tray. We slice the fillet and all agree that this is a perfectly cooked medium-rare fillet of beef, and we discuss what it felt like when it came out of the oven, whose arse cheek or breast or bicep or thumb it most recalls.

We do this because it's important to try and remember that touch. The steak is hot when it comes out of the oven and it always looks more cooked than it is, because we've seared it until the flesh has caramelised in the cast-iron pan, but the touch: cooking steak to order is all about the touch. Whatever else I learned in hospitality, however long it took me to sort my life out, I learned how to cook steak properly, and how touch was everything.

Jim Hearn is a scriptwriter and chef. He worked on the screenplay for *Chopper* and an adaptation of Andrew McGahan's novel *Last Drinks*. He is writing a book about his experiences in the hospitality industry.

Food security in the Arctic

Finding the key to survival

Annmaree O'Keeffe & Chester Reimer

IN 1847, four years after being stood down as lieutenant governor of Tasmania, Sir John Franklin died at the other end of the earth trying to find the Arctic's fabled North-West Passage. He and his crew were too proud to ask the local Inuit communities for advice. Like the Norse in Greenland some five hundred years earlier, they died partly because they would not accept that an indigenous people held the key to survival in the Arctic.

By the time Franklin had been immortalised for his intrepid but failed expedition, the Inuit had been living in their frozen homeland for millennia. They survived where many others perished because they adapted their diet and customs, their culture and even their language to this frozen and hostile yet resource-rich wilderness.

Inuit continue to survive in their northern homeland, but it has not been easy. Up until the middle of the last century, hunger and starvation were not uncommon during the long, dark winters, particularly for the more remote communities. As the Arctic became more accessible, after World War II, indigenous communities had greater access to reliable food supplies from the south, and the spectre of winter hunger started to disappear. But in its place have come other, more complex threats to Inuit food security and wellbeing, the product of decisions made far from the Arctic. The result is economic vulnerability, contaminated food and changes to the movement of the Arctic land and sea mammals – the source of traditional Inuit food.

While the Inuit survived independently in the Arctic for thousands of years, observing a passing parade of explorers and whalers, well-meaning missionaries and would-be settlers, their survival is now precariously linked to an outside world which has long held romantic images of – and grand ambitions for – the Arctic, and whose actions erode the means of Inuit survival. The lure of the Arctic persisted because the region is so close to the old and new powers of the nineteenth and twentieth centuries, but remained inaccessible due to its climate. Franklin and many others tried to find a passage through the region to save time transporting goods between the major trading nations – an ambition that has now almost been realised, not because of navigational success but because of the melting Arctic ice.

THERE ARE 160,000 Inuit across the Arctic: 48,000 in Alaska, 55,000 in Canada, the same again in Greenland and about 2,000 in Russia (Chukotka). Their homeland spreads from Greenland across the Arctic stretches of North America and over the Bering Strait to the eastern tip of Russia. They are a diverse but cohesive group working for unity despite different political and economic circumstances. In 1977, they formally expressed this unity by establishing the Inuit Circumpolar Council (ICC), a powerful player in international negotiations over matters affecting the Inuit.

To understand the significance of food security for the Inuit, we must recognise that food security isn't simply reliable access to nutritious food. It is linked to climate change, wildlife management, pollution and economic vulnerability – and to cultural security. According to 'The Legal Protection of Subsistence: A Prerequisite of Food Security for the Inuit of Alaska' (*Alaska Law Review*, 2007): 'food security goes beyond the mere satisfaction of physical needs – it integrates the social and cultural symbolism of food, which determines *what* food is and *which* foods are appropriate for human consumption…Inuit still partly derive their self-worth, individually and collectively, from traditions associated with hunting, fishing, and gathering. More than a mere means of obtaining the foodstuffs required for physical survival, these practices represent an important aspect of community integration. Activities related to subsistence represent an important foundation for the social and economic organization of Inuit communities.'

Remoteness, limited infrastructure, difficult climatic conditions and fluctuating prices for food commodities and oil combine to make food expensive and, at times, its quality questionable for many Inuit communities. Canada's Department of Indian and Northern Affairs conducted a survey of isolated northern communities in 2006 and 2007 to gauge the weekly cost of a nutritious diet for a family of four. The results showed that it was at least twice the amount that a family living in southern Canada would have to pay.

Families living in remote communities also have other high costs: fuel and related transportation expenses essential for hunting activities. At the same time, low income levels, limited job opportunities and widespread dependence on welfare payments mean there is insufficient money to cover basic food needs.

Nain, in Canada's north, encapsulates the challenges confronting small and remote settlements in the Arctic. With a population of just over 1100, Nain is the administrative capital of Nunatsiavut, the smallest of the four Canadian Inuit regions. The settlement was originally established by Moravian missionaries in 1771, attracting some of the nomadic Inuit to settle there, and is now the northernmost town of any size in the Canadian province of Newfoundland and Labrador.

There are no roads leading anywhere from Nain and the only year-round access to the town is by air. But there's no ground support, so landing is by visible aids only and when the rocky hill overlooking the town is shrouded in cloud, which is often, no plane can take off or land. For a couple of months in summer after the sea ice has melted, ships dock, providing employment to those who unload the cargo. There's a labradorite mine nearby which employs maybe fifty or so people; a fish-processing plant in previous years employed another fifty for three to six weeks, but even that looks increasingly unlikely – it's just not economical for the fishery owners.

Economic vulnerability is exacerbated by the environment. It costs C$800 a month to heat a house, so families move in together to save costs. Fuel is so expensive that it limits traditional hunting expeditions, which heightens reliance on store-bought foods. Alcohol consumption is the second highest in the province, second only to the capital, St John's, in Newfoundland. With alcoholism comes a raft of social and personal problems. But the basic services are pretty good for a town this size – the health clinic is more than adequate and its local service is supplemented with telemedicine facilities; there's a high school and being the administrative centre for Nunatsiavut brings kudos and a little employment. Voisey Bay, a nickel mine south-west of Nain, provides some employment while the Nunatsiavut government receives 5 percent of provincial revenues from the mine.

IN MAY 2008, the US Fish and Wildlife Service announced its decision to list the polar bear as a threatened species 'throughout its range'. This decision was the outcome of a process which had seen environmental groups square off against the Inuit. Partway through the process, three American-based environment organisations, including Greenpeace, had instigated legal proceedings against the American Government for the delay in making a decision on whether to list the polar bear.

The polar bear is a cuddly global symbol of the threat posed by human-induced climate change. For the Inuit and anyone else who lives close to them, though, the polar bear is certainly not cuddly and it represents an important food source which local communities insist they manage sustainably.

While the listing is a domestic American matter, it has ramifications across the Inuit homeland exacerbated by the US proposal in 2008 that the Convention on the International Trade of Endangered Species upgrade the status of the polar bear to 'most endangered'. The Americans' move has been celebrated by environmental groups as a positive step in their campaign to pressure the nation's government into changing its position on climate change. For the Inuit, this represents a further erosion of their capacity to manage the resources that have sustained them through the centuries. The Inuit had experienced this kind of foreign pressure before. In the 1980s, following a massive international campaign spearheaded by Greenpeace, the European Union banned the import of certain seal products. While viewed internationally as a victory for animal rights, and not directly aimed at Inuit hunting methods, its effect on indigenous communities in Greenland and northern Canada was disastrous. To its credit, Greenpeace, when made aware of the consequences of its action, apologised – but by then the suicide rate among young Inuit men had already surged.

IN 1962, RACHEL Carson published her groundbreaking book *Silent Spring*, which described for the first time the detrimental effects of organochlorine pesticides on the environment. Despite the grimness of her findings, there was some good news: the Arctic had escaped contamination. By the 1980s, this was no longer the case. Persistent organic pollutants (POPs), heavy metals and radionuclides had reached very high levels in the Arctic ecosystems. This surprised scientists, because the substances were produced and used in industrial regions far from the Arctic – in Europe, Russia, Canada and the US. The Inuit's once-pristine diet of marine mammals was now among the most contaminated in the world.

While POPs, which include chemicals such as DDT, are not manufactured or used in the Arctic, they end up and are trapped there, simply because it is colder than elsewhere. And POPs are lipophilic: they attach themselves to fatty tissues. Because the Inuit depend on whale blubber, seal meat and various oils, and because POPs quickly bio-accumulate up the Arctic food chain, Inuit women have seven to eight times more of these carcinogens in their breast milk than do women in Toronto or Sydney.

Thus commenced two decades of international lobbying by the Inuit – mostly through the ICC – to make the world aware of what was happening to their principal sources of food and to support other international efforts to combat pollution. One of the most important outcomes of this lobbying was the Stockholm Convention on Persistent Organic Pollutants, which 114 countries signed in 2001 and which came into force in 2004, the same year that Australia ratified it.

The global efforts have had some success and recent reports point to a slight decline in the levels of some contaminants in the Arctic environment. But studies are now showing that a number of contaminants, such as mercury and newly emerging POPs, are not decreasing. As temperatures rise in the Arctic, ice- and snow-locked mercury is being released and finding its way into the food web.

The choice for Inuit has been to eat and be damned, or not to eat and still be damned, because the marine-based diet has been steadily replaced by less-nutritious store-bought commodities which are increasingly contributing to health problems such as obesity, heart disease and diabetes. Health authorities around the Arctic support continued consumption of the marine diet, but with a greater emphasis on those animals further down the food chain, such as fish.

ACCORDING TO THE United Nations Environment Program, the effects of climate change and the consequential effects on snow and ice are already being felt in small communities throughout the circumpolar north. Among the most significant, according to a 2007 UNEP report, are 'health and nutritional concerns (related to the availability of country food) associated with changes in the abundance and migratory patterns of subsistence resources'. This is reinforced by the observations of traditional Inuit hunters who have been reporting for two decades that major changes are occurring in the migratory habits of the marine and land mammals which form the bulk of traditional Inuit diets. Inuit hunters have adapted to these changes, but it now takes more time and more fuel to cover the distances needed to find traditional food sources. As ice coverage becomes more fragile and unpredictable, the risk of more hunting accidents grows.

Ironically, as food security for the Inuit is increasingly linked to the wider world, the future of the world's food supply has been entrusted to the security of the Arctic. In 2008, the Global Crop Trust Fund announced the establishment in Norway's Arctic regions of the Svalbard Global Seed Vault – the only global back-up seed storage facility, popularly known as the Doomsday Vault – as insurance against the destruction of the world's crop seeds. Thanks to the Arctic permafrost, we can be sure that the seeds will be conserved, even in the event of electricity failure. But will the Inuit also be assured of long-term food security?

References available at www.griffithreview.com

Annmaree O'Keeffe has worked in government and international aid organisations since 1980 in a number of developing countries, including Papua New Guinea, where she headed the Australian Government's aid program, and Nepal, where she was Australia's ambassador. She was Australia's first ambassador for HIV/AIDS. More recently, she has been working with the Inuit Circumpolar Council.

Chester Reimer has represented Inuit at the United Nations, the Arctic Council and other international forums, and has a long history of working with the Inuit Circumpolar Council and the Arctic Council's Indigenous Peoples' Secretariat. He heads CRCI, a Canadian-based consultancy.

A safe house for humanity

MARI TEFRE

The world's seed collections are vulnerable to
war, natural catastrophes and, more routinely
but no less damagingly, poor management,
lack of funding and equipment failure.
Unique varieties of our most important crops
are lost whenever disaster strikes.

The Global Seed Vault answers a call from
the international community to assure the
safety of the world's crop diversity. Securing
duplicates of all seed collections in an
international facility is an insurance policy for
the world's food supply.

The vault is in a mountainside near
Longyearbyen, a village on the Svalbard
archipelago nearly a thousand kilometres north
of mainland Norway. For nearly four months
a year the islands are enveloped in darkness.
Permafrost and thick rock ensure that, even
without electricity, seed samples remain frozen.

Images © Mari Tefre. Courtesy of Global Crop Diversity Trust: A foundation for food security. www.croptrust.org

ESSAY

Food in the age of unsettlement

Finding new ways to feed the hungry

Tony Fry

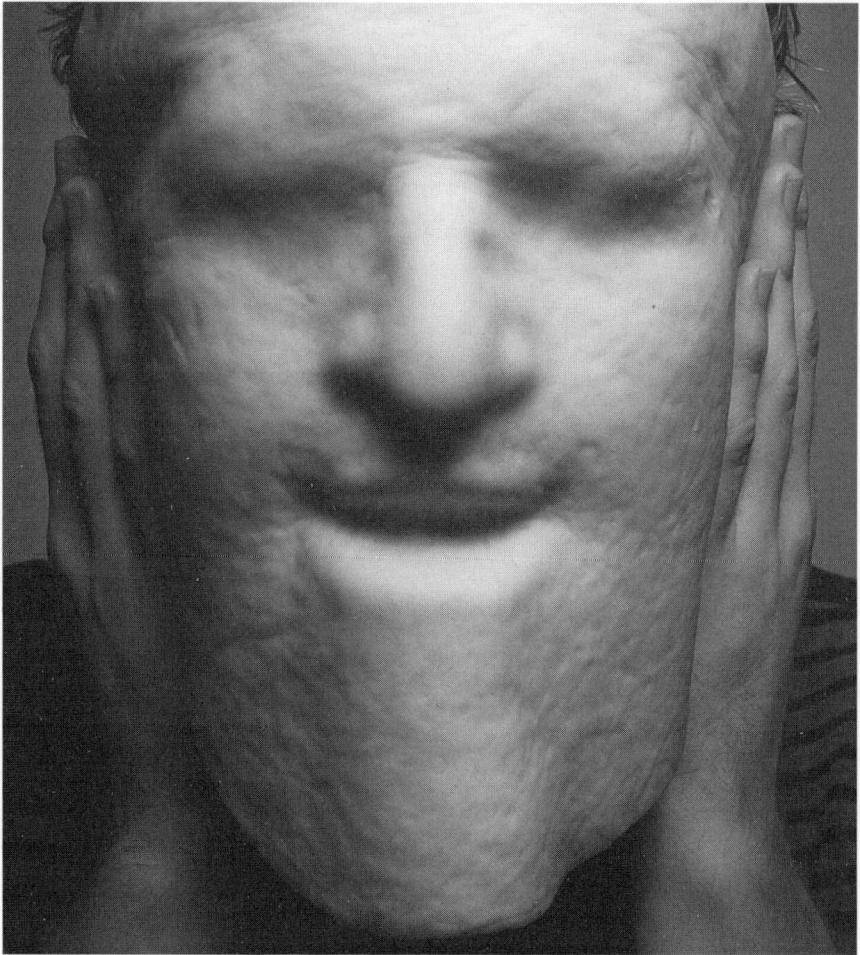

I have three personas: designer-educator, sustainment theorist and forest farmer. The farmer came first. My grandfather was a market gardener who taught me to grow vegetables before I went to school. More than half a century later, I'm still doing it. The garden at our farm on the ranges of southern Queensland supplies the kitchen all year round and produces commercial quantities of sweet potatoes, chillies and wild hibiscus.

The wild hibiscus is a thread connecting much of what I have to say here. It comes from Africa, where it is known as *karkade*. It is a deep-rooted plant able to survive with little rain. In countries such as Sudan, it's a staple: the deep magenta calyx, the bit that looks deceptively like a flower, is dried to make a soft drink and tea high in vitamin C; the leaves are used as a salad base; the seeds are ground for flour; the stems are dried and used as kindling. It is sustainability realised.

Around the world, wild hibiscus is used in herbal teas, and many Australians are familiar with it as rosella jam. More recently, the calyx has been bottled in syrup and used, like cassis, to turn champagne pink. Most of our crop goes to market, but we also supply the Sudanese community in Toowoomba with dried calyx.

What I teach, the things I design, what I think about and where I work: all are connected, and all address sustainability.

TO BE SUSTAINED, to sustain others and the world around us depends on creating and maintaining sources of nourishment. For many people in the developed world, this is assumed. For most people in Timor-Leste and the rest of the developing world, gaining nourishment – by growing it, trying to find work to earn money to buy it, or simply stealing it – is a struggle every single day. Somehow, global food problems never get sufficient recognition. As soon as a big issue comes along – climate change, the global economic crisis or the so-called war on terror – it is pushed into the background, yet food security is embedded in these and many other issues.

It has been estimated by Josette Shearan, the head of the UN World Food Program, that to feed the projected global population of more than nine billion by 2050, food production will have to double. Climate change and increased global tensions are likely to produce huge increases in the number of displaced people searching for food. The food riots in more than forty countries in 2008 were a glimpse of a likely future.

By 2009, according to United Nations figures, more than a billion people were hungry – more than ever before – with greater than two-thirds of them living in the Asia-Pacific. Despite epidemics of obesity in many countries, the annual global per-capita intake of food is decreasing. The more agriculture is 'liberalised' in a 'free-market' system, the more 'high-return' crops are grown, the more food prices rise and the more people go hungry. When you 'live' on one dollar a day or less, as most of these hungry people do, even a small price increase can make the difference between a meal a day and a meal every other day.

Working in Timor-Leste, you see the reality of these facts in so many ways. You encounter so many malnourished people on the street every day; note the arrival of the Red Cross hospital ships sent, in part, as a response to the nation's enormous infant mortality rate (not least due to the unavailability of food, especially in some of the rural areas); and you are aware of families getting up in the early hours of the morning to work to earn a few additional cents before starting their lowly paid 'day job'. Working in Timor-Leste, I see these problems compound: irregular meals, combined with a poor diet, result in a weak immune system and increased susceptibility to disease.

That a nation may be rapidly industrialising does not mean all, or even most, of its population is well fed. Vandana Shiva – the physicist, author, director of the Research Foundation on Science, Technology and Ecology, vice-president of Slow Food International and chair of an international commission on the future of food – notes that there are more hungry people in India than in Africa. Although India has enjoyed more than 9 per cent growth in GDP, half its children have severe malnutrition. About half these children will die before they turn five. The global financial crisis further reduced incomes and increased unemployment. Jacques Diouf, the director-general of the UN's Food and Agriculture Organization, estimated that it increased the number of hungry people by about a hundred million. Echoing Malthus, he sees in food shortages a 'serious risk for world peace and security'.

THERE IS NOT just lack of food, but insufficient money to buy it. Hunger and poverty cannot be separated. As the anthropologist Marshall Sahlins wrote nearly forty years ago: 'Poverty is not a certain small amount of goods, nor is it just a relation between means and ends; above all it is a relation between people. Poverty is a social status. As such it is the invention of civilisation.'

World food prices are likely to continue to rise, and to take a terrible toll. It is estimated that in the three years to 2006, two hundred thousand Indian farmers took their own lives (some by drinking insecticide) as high seed prices, indebtedness and political pressures – driven by market liberalisation, and the need for greater productivity and profit – pushed them beyond despair. The farmers borrowed money to buy expensive high-yield seed, rainfall was low and crops failed. Many farmers lost their land as a result, but not their debt.

Many poor nations prioritise earning foreign exchange over responding to local food needs. Negative economic and ecological forces are set to converge, and inequity is expected to worsen. There is a crisis of cultivation, especially in the developed world. The cost of food production continues to increase, while investment in research and development declines and agricultural profit margins shrink. This drives down wages in developing countries where industrialised agribusinesses prevails – and climate change is likely to worsen the situation.

Climate change isn't simply linear, cause and effect; it is a complex set of relational processes. The atmospheric life of greenhouse gases is long – even if emissions are significantly reduced in this century, the problem will remain for a long time. Deep ocean temperatures play a major role in the global climate system, acting as a thermostat, and influencing ocean currents and weather patterns, and would take a long time to change. Global warming is reducing soil moisture; increasing evaporation means that more rain is required to recharge the soil to support viable levels for cropping, and deep-rooted plants succumb to stress and become more vulnerable.

Traditional and sustainable farming practices retain soil moisture by maintaining the soil in good condition and reducing the exposure of broken soil to the sun. Industrial agriculture demands increased productivity, and relies on irrigation to retain soil moisture. As a result, it can require up to ten times more water to produce the same amount of food as ecological farming practices. Agricultural irrigation demands large dams, alteration of river flows and groundwater mining, with adverse ecological consequences.

The problems of industrial agriculture extend beyond its impact on the land. Dependence on large quantities of fertiliser and an extensive transport and distribution system requires a great deal of fossil-fuel-based energy. Agriculture is the second-biggest carbon emitter after road transport, representing between a quarter and a third of greenhouse-gas emissions. The UN estimates that a fifth of this can be attributed to the emissions required to feed and transport livestock, which consume almost half the world's grain production. As the American food writer Michael Pollan pointed out in his open letter to 'farmer in chief' and President-Elect Barack Obama, in the *New York Times* of 12 October 2008: 'When we eat from the industrial-food system, we are eating oil and spewing greenhouse gases. This state of affairs appears all the more absurd when you recall that every calorie we eat is ultimately the product of photosynthesis – a process based on making food energy from sunshine.'

THE REALITY OF this in a place like Timor-Leste is stark. The urban fabric is damaged and neglected, as is the infrastructure. Many power lines were removed as the Indonesians withdrew, and people reverted to cooking with wood. Poverty made this fuel the sole option. Consequently, the trees around towns and cities are stripped, some die and soil erosion sets in, topsoil is lost and the ability to grow food diminishes. Unsustainable agricultural practices are deeply embedded, which makes change slow and even generational.

Food has to be localised wherever possible, and economists need to factor in environmental costs along with traditional profits. As Vandana Shiva points out, 'more than 25 per cent of climate instability is being caused by unsustainable farming'. When transportation and related industries are included, the figure increases to 35 or 40 per cent. Industrial agriculture 'displaces small peasants, creates poverty and bad food'.

Ecologically responsive farming can address emissions, poverty and food quality, and provide significant employment opportunities – food production begs to become a major sustainable industry, including in nations where many people have abandoned the land. This should not mean a return to basic subsistence farming, but rather the co-operative development of sustainability by appropriate rotation and high-organic-matter farming methods that deliver a liveable income and quality food. The bias will need to shift from food processing to primary food production.

Although cities cover a mere 2 per cent of the planet's surface, they command nearly three-quarters of its extracted resources and are where most of the world's people now live. Almost all cities could produce much more of the food their residents need, better manage and utilise their organic waste and storm water, and more productively manage vegetation. According to the UN Development Program, in the past twelve years urban food production has more than doubled, and now accounts for almost a third of the world's food production. Necessity, not choice, has driven this. It is a response to inadequate, unreliable and irregular access, a lack of purchasing power or employment.

The situation is global: in greater Bangkok, nearly two-thirds of the land mass is under cultivation, with almost three-quarters of the population involved in growing food, mostly in their spare time. Figures are similar in Moscow and in Dar es Salaam, Tanzania. The Food and Agriculture Organization estimates that two-thirds of urban and peri-urban households in developing countries are involved in urban agriculture, a trend that benefits women and children, saving time and money and improving child health.

Timor-Leste is still underdeveloped – there are scraggy chickens, pigs and goats everywhere. Some urban food crops are grown, but not many: much more could be produced. Urban agriculture needs to be a part of the nation's urban regeneration.

WHILE URBAN FOOD production in the developed world tends to be dominated by 'alternative culture', it is not a marginal issue for most of the world's people. Urban food production has been the subject of a world summit, global forums, numerous international round tables and declarations, as well as countless events concerned with organic waste and water usage, resources and conservation. It is supported by government and non-government organisations and research centres worldwide, has a considerable body of literature, and is serviced by a number of dedicated websites and publications – not least the substantial Netherlands-based *Urban Agriculture Magazine*, published three times a year in six languages.

In times of conflict, when a nation's food supply and security are under threat or restricted, food production, particularly urban food production, increases significantly. During World War II, food grown in backyards, public parks and urban wastelands provided much of the food for the population at home in Britain, the US, Australia and other countries. If that method were applied today, it would drive an enormous amount of land use toward horticultural innovation. Vertical farming, which combines intensive methods such as hydroponics with the surface-area cultivation of certain crops on multistorey buildings, is already underway.

Initiatives like this could make cities better, healthier and more sustainable places to live and work. Intensive urban agriculture is environmentally more sustainable, and can lead to positive urban development: it can generate small and large businesses, contribute to urban design, assist waste management, consolidate and build community, provide training and work opportunities for the unemployed, and improve health through increased exercise and better diet. It can also aid climate-change adaptation – green roofs reduce the heating of thermal mass, and a cooler city means a smaller energy load.

Urban agriculture can also contribute to developing the knowledge and skills that will assist in coping with unsettlement: a potentially powerful learning environment for the urban population from infancy to old age. (It would, though, need proper regulation and micro-management to prevent 'green chaos' and cattle roaming the streets.) In the coming age of unsettlement, millions of people in cities around the world will be exposed to civil disorder and looking for food. If the patterns associated with climate change continue, where and how people work and live in the city will alter radically. Ultimately, the degree of destabilisation will be determined by how we plan to adapt, and how we act on those plans.

Tony Fry is adjunct professor and convener of the Design Futures Program at the Queensland College of Art, and a director of the sustainability-consultancy firm TeamDES. He also leads a Griffith University/AusAID-funded project in Timor-Leste.

MEMOIR

Hunting with the boys

In search of bears

Wayne McLennan

AS soon as we heard the growl, Hans released the safety on his rifle and, in the same motion, swung it from its cradled position across his chest and aimed toward the bushes. 'Step back...slowly,' he cautioned. We took two backward paces. Behind us, the creek bubbled towards the inlet. Sandbars and shallows had slowed its flow, giving the salmon, who fought frantically to reach their spawning grounds further up river, a brief respite.

'Let's go,' he said, after only a moment, and we moved into the bushes. The bear had left. Only a swathe of flattened grass, a blob of excrement and a smell that reminded me of wet pine needles marked its passing.

We re-entered the creek on the far side of the bushes. The water was thigh-deep, but our waders kept us dry until we reached the shallows. Salmon lay about us, dead and dying, flaking apart in the current; others burst between our legs and around us as thickly as spring pollen. We moved upstream tenderly, measuring every step.

A sow and two large cubs were fishing in the reeds ahead, pawing and slapping the water. With a drop of his hand, Hans signalled us to squat, while he flicked on his lighter, checking the wind direction. When he was confident that they could not pick up our scent, he motioned us out of the water and on to a trail made by another bear, at another time. We were only here to shoot big males.

Further upstream, we moved away from the river, climbed a steep hill held together by the roots of an old hemlock, and found a log that gave us an unobstructed view over shallow rapids. A likely bear fishing spot. We nestled with our backs against the fallen log and waited. Hans pulled out a packet of Copenhagen and settled a pinch of the tobacco behind his bottom teeth. John chewed on a wad of Redman, expectorating a blood-coloured stream every few seconds.

Seagulls rode on the backs of swimming salmon, pecking them to death. Bald eagles soared lazily, seemingly disinterested in the feast below. A waterfall above the rapids roared and foamed, wetting and chilling the air, making me shiver regularly.

PREVIOUS PAGE: *Chef Sean Connolly.* Photographer: Jack Atley
from the *Sydney Morning Herald* Shoot the Chef! 2009 photographic competition.

When we got back to the beach, the tide was creeping stealthily toward our launch. Another fifteen minutes and we would have been stranded. Hans and John unloaded their rifles, and Hans pulled on the line that anchored the launch to the shore, hand over hand, coiling the rope meticulously as it fell toward the ground.

Nobody spoke as we motored back to the boat. Hans stood at the wheel, steering us through a chop that had grown to three feet. John sat on the middle seat, stains of dried red spittle at his feet. I hunkered down in the bow, trying to stay out of the wind and spray, ready to make the launch fast when we arrived.

Hans suddenly threw the skiff into an arc, and when I looked carefully over the gunnels, I could see plumes of water shooting into the air. We got within thirty metres of the humpback whales before they sounded, and then they were gone.

IT HAD BEEN twenty-five years since I had last seen Hans. He'd hardly changed. Still all muscle, hair still blond, only grey five-day growth giving an indication of his age. He was prosperous now, an outfitter and hunting guide. His company, South East Alaskan Guiding, was one of the most respected in the country. He promised his clients a good chance of killing brown bear and mountain goat, the two most sought-after trophies in American big-game hunting, the two most demanding and dangerous. He had an impressive success rate.

Hans transported his clients to the hunting grounds of Admiralty Island, forty miles south of Juneau, Alaska, on his fifty-one-foot long-line fishing trawler, the *Northern Star*. I recognised her at first sight. Not new, but immaculately cared for. She looked like an Eton boy among urchins.

The *Northern Star* slept five comfortably; it had a stove, oven, fridge, hot shower, toilet and clothes dryer. A generator powered the electrics. But it was also a working fishing boat, with a satellite-navigation system, depth finder, radio and autopilot.

The *Alaska Shark*, the boat I worked on so many years ago, had a compass, charts, ruler and pencil. Landmarks were used to avoid rocks, and you had to steer the boat yourself and shit over the side, or in a bucket when the weather was bad. Even on Hans's boat we still shat over the side…old habits die hard.

Hans had invited me to join him on a bear-hunting trip. You couldn't turn an offer like that down. Or the chance to see an old mate.

JOHN ALWAYS CLIMBED awkwardly out of the launch and onto the boat. Bad knees. He never complained, even when he was in a lot of pain, which he was when his pain tablets started wearing off, or when Hans pushed him to the limit. Hans had no tolerance for weakness.

It was John's fourth hunt with Hans. It wasn't that Hans could not find him a big bear – just that John wanted the biggest and was prepared to wait. He had already spent $50,000, but it had nothing to do with money.

When I asked him whether he thought he was ever going to get his bear, he laughed and answered in his syrupy Arkansas drawl, 'It's an island – he ain't going anywhere.'

We had been up since 4.30 and had eaten only a few mouthfuls of cereal. Hans had already begun frying bacon bits, diced potato and a small hill of onions before John and I were out of our waders. I poured coffee, before walking aft to light a cigar. Protected from the wind by a scimitar of land, the water around the boat was as flat as a skimming stone. The sun had burned away the early morning fog, letting the sky mingle with the water, in a lapis lazuli of blues and greens. Outside our cove, the waves rushed by in low walls of grey and foamy white.

JOHN IS WHAT many Americans call a good ol' boy, some a cracker. He lives in a town that stayed segregated until the mid-1960s: the picture theatre, bowling alley, swimming baths, drinking fountains. 'Even when they took the signs down that separated humanity, people still stayed on the same side of the picture theatre they always had, used the alleys that they always bowled on, drank out of their own fountains,' John told me with considerable consternation.

John likes to eat grits and gravy; he is a member of a private business club, a town leader, a former marine, a Vietnam veteran. He lives in a large house filled with the grim detritus of his hunts throughout the world, calls black Americans 'coloured' and runs a well-digging company started by his 'daddy' that he took to new fiscal heights. At one time, John employed almost exclusively black Americans; now he hires only Mexicans. 'They're better workers – coloureds don't want to work anymore.'

I was tempted to question this generalisation, but he had worked himself from nothing to something without an education, and who was I to judge?

'Whites don't want to work either,' he suddenly added, expectorating pink into a plastic bottle that he used when sitting at the eating table.

IT WAS OUR habit to sleep after eating, rising after a couple of hours to eat again, usually soup or cold cuts, and prepare for the night's hunt. Hans and John would check their rifles, working the bolt actions, rubbing them cautiously with lightly oiled rags, caressing them tenderly. Then we would dress. My attire consisted of three sets of long underwear, three pairs of socks, pants and thigh-high waders held up with string tied to my belt, stout gloves, woollen cap and rain jacket. I looked more like a butter bean than a hunter, but I thought that better than being cold. John and Hans always dressed much more lightly. John always wore camouflage.

After we dressed, the necessities were assembled: binoculars, the heavy night-vision scope, headlamps, knives. The launch was pumped free of any water, and the outboard motor's tank filled with gasoline. Bullets and tobacco were essential.

'Did I tell you I have a business in India?'

'No, John,' I answered, as we sped out into open water. The passage had calmed, and we raced over waves so gentle they resembled freshly whipped cake batter.

'I have a foundry there.'

I waited, understanding that timing was important in John's stories.

'I met an Indian man on a plane, trusted him, and we took it from there.'

'Was he trustworthy?'

'No, but my new partner is!'

Hans had decided to hunt Moon Cove. We had been there before; there was no cover and no high ground to get out of harm's way, if it came our way. We were forced to crouch among the tidal rocks choked with mussels, amid salmon corpses, next to a creek that rushed violently into the sea.

Before we landed, Hans motored slowly back and forth, scanning the beach with binoculars, looking for the best place to land without disturbing the sows and cubs that were already feeding along the creek's banks. It was low tide, cloudless, and a full moon was expected. It would be a long and bitterly cold night.

WE NESTLED INTO the rocks, rubbing our backs against the mussels, trying to find angles that would give us comfort. Across the creek, sows were feeding with their young. It was the gloaming and we waited for the moon, watching their silhouettes dip into the water. Hans pulled out his night-vision scope, set it carefully on a rock and scanned both sides of the creek. The moon came up and the cold set in. John and I squirmed and wiggled, trying to stay warm; Hans remained as still as the rock he was squatting on.

'Never did tell you about the tigers, did I?'

'Give me some of your Red Indian,' I answered, peering warily into the dark for large shapes, before stuffing a ball of tobacco in my mouth.

'Them ol' boys were man-eaters.'

I spat thickly, the red liquid running down my chin. Nausea came on almost immediately; my head began to spin. The tobacco was too strong, but I had set my mind to continue chewing.

'We captured two of them, and we was walking through the jungle looking for the third when he came charging out of the undergrowth.'

'You were on a trophy hunt?'

'No,' answered John, mildly annoyed that I had interrupted his story. 'I was doing a favour for the *National Geographic*. They wanted to film the capture of these old tigers that had been eating some of the villagers, and they asked me to go along to protect their people: two ol' girls, both about twenty, and the cameraman.'

I nodded, raking the tobacco off my gums, trying not to swallow any of the amalgam left wallowing like swamp water in my mouth.

'That ol' tiger came charging straight out of the bush. It was only 'bout ten feet away when I shot him. Them armed guards they sent with us already dropped their rifles and ran. So did the cameraman. Can't blame him though…I suppose.'

'A man-eater?'

'That's what they say.'

'What about the assistants?

'Those two ol' girls couldn't even talk they were so scared.'

THE NIGHT WAS so still that a whisper sounded like a shout. Hans threw us an annoyed look and then changed position, turning to face across the creek. A new bear had joined the others.

The gender of a bear is extremely difficult to determine, and almost impossible in the dark. Hans was concentrating on the new arrival, changing the focus of his night-vision binoculars, adjusting the range, repeating the procedure, and then repeating it again.

'Can't see their balls,' John whispered to me. 'They don't hang down. A sow looks just like a boar. Usually a lot meaner, though.'

Another bear moved in from our left. Hans and John noticed her only when she was almost on top of us. I became aware when Hans and John swung around to face her. She caught our scent and moved back over the tidal rocks and along the beach.

I was caught unawares again, when a full butter-coloured moon rose over the island's high forests and lit up the beach and creek like church candlelight.

Noiselessly, a new bear broke out of the forest. It was long, lean, and Hans needed less time in the better light to see that it was a boar. It played at the water's edge on our side of the creek, rolling rocks, splashing the foam with its paw, lifting a salmon out of the water only to let it drop back again.

'Too small,' Hans growled, and turned his attention back across the creek.

'Let's get a closer look.' They moved, stopping behind a large boulder directly across from the bear. Hans set his night-vision scope on the rock and began his study. Small dark clouds appeared, blocking out the light, drifting away again, as if playing hide and seek with the moon.

I remained squatting behind my rock, watching my bear with fascination, feeling very much alone. Without warning, the bear sloshed into the creek, pushing rocks playfully in front of him. I could see him clearly. He was lean but heavily muscled, and I was sure he was big enough for John.

The bear suddenly turned, wadding slowly towards me, breaking the water like a steaming boat. I moved behind another boulder, picked up a rock, and cursed that I was not armed. Hans and John crouched thirty feet away, their backs towards me. I waited. It got to within ten metres before it turned, grunted and disappeared into the shadows on the other side of the creek.

Hans and John returned with the news that their bear was a sow.

'What's wrong with your shoulder?' I asked, when I noticed Hans rubbing it.

'Popped it.'

Hans began to study my bear again with renewed interest when I told him how big I thought it was in silhouette. He watched for an hour, straining at times to catch sight of him in the shadow of the woods, before again proclaiming him too small. 'If you had seen him side on,' I whined – but Hans couldn't be convinced.

We waited until the tide arrived, bubbling about us, carrying with it weeds and more dead salmon, and then we stepped cautiously over the partially submerged rocks toward the anchored boat.

A medium-sized bear came out of the woods, grunting, belligerent. 'Side by side,' Hans ordered. To a bear, we would appear as one, and therefore large and possibly dangerous. The bear hesitated, grunted and grunted again, before ambling casually away. He remained a short distance from us, grazing on salmon, lifting his head intermittently, challengingly.

'That's one bear that ain't scared of us,' Hans remarked, lifting his shoulder horizontally as if it were a hinge that needed adjusting.

'Rocks are slicker than owl shit,' John said, seeming not to have heard him, concentrating instead on his struggle to scramble into the boat. When we started the motor, the bear raised his head again, and then we lost him against the darkness of the forest.

'In the morning we hunt Black Deer Creek.'

John nodded before finishing his whisky in one gulp and heading below to sleep. A light breeze rocked the *Northern Star* like a cradle.

THE FOOTPRINTS ON the beach were eight inches across and very deep. 'He was down here last night. He's probably not more than a few hundred yards away, sleeping the food off,' Hans said.

'How big?' I asked.

'That's a big bear – got to be more than nine and a half foot.'

'Come back tonight?' John asked.

'Or we can hunt him now. Be in his territory, though. Have to be real quiet and real careful.'

It was steep, and John's knees were no good. I thought he would decline, but he just nodded. We moved off, step by measured step.

After we climbed off the beach, we lost the sun and the trail became slippery. Moss enveloped every fallen log; vines crossed the trail like the rungs of an endless ladder. Each time we broke a twig underfoot, Hans would turn and glare.

The bear was easy to follow in the beginning. It had skidded down to the beach, scraping the vegetation away, leaving a wide muddy trail. As we climbed, it became more indistinct; multiple tracks appeared. Hans followed the freshest with little trouble, checking every mossy glade, squatting, waiting, listening, peering behind every broken spruce, hemlock. Shadows jumped out at us. The woods were so thick and tangled, he could be anywhere.

As we crept higher, the footing became more difficult. John was breathing heavily. He had stripped off his jacket; sweat dribbled across his face, which had turned as red as a ripe peach and disappeared into his beard.

Hans held up his hand and we stopped. He pointed ahead and above us. I was ordered behind a tree and told to keep watch to our left. John shouldered his rifle.

'If he comes, he'll come fast, so be ready to warn us.' I nodded.

'Bite your thumb,' John said, almost in a whisper.

Hans picked a blade of grass, placed it between his lips and imitated the distress call of a black-tailed deer. A high-pitched whistle, desperate and morose.

Nothing! He repeated the call. We waited, seconds seeming like minutes.

'Let's go.' Hans indicated with a hand movement.

We climbed onwards to the top of the ridge: only 1,500 feet, but it had taken us two hours.

Hans stripped off his pack; we did the same and followed him along the ridge, creeping, bent double. Suddenly Hans straightened and I knew we had lost the bear. We walked another twenty metres to his sleeping bed at the base of a tall spruce, a circle of dirt littered with branches and faded needles. The bear had given himself a view over the incline we had just climbed. We could still smell him.

'Probably watching us the whole time,' Hans said admiringly, working his shoulder, trying to lift his arm a little higher than the last time.

'What did you mean by "bite your thumb"?' I asked John as we walked back to retrieve our packs.

'It keeps ya concentrated. An ol' boy told me that when I was a kid. It's the pain that does it.'

When we got back to the beach, John began to knead his knees, rolling the caps around with the palm of his hand as if they were rubber balls.

'Disappointed?'

John clamped his mouth against his pain and began to shake his head. 'It's hunting; otherwise it would just be killing.'

Wayne McLennan is the author of *Rowing to Alaska and Other True Stories* and *Tent Boxing: An Australian Journey*, both published by Granta Books. His pieces in *Griffith REVIEW* include 'Night at the fights', 'Meat' and 'My banker'. This memoir is an extract from a longer story, 'Hans's bears', written with the assistance of a grant from the Australia Council for the Arts.

ESSAY

Fishing like there's no tomorrow

Behind our seafood feast

David Ritter

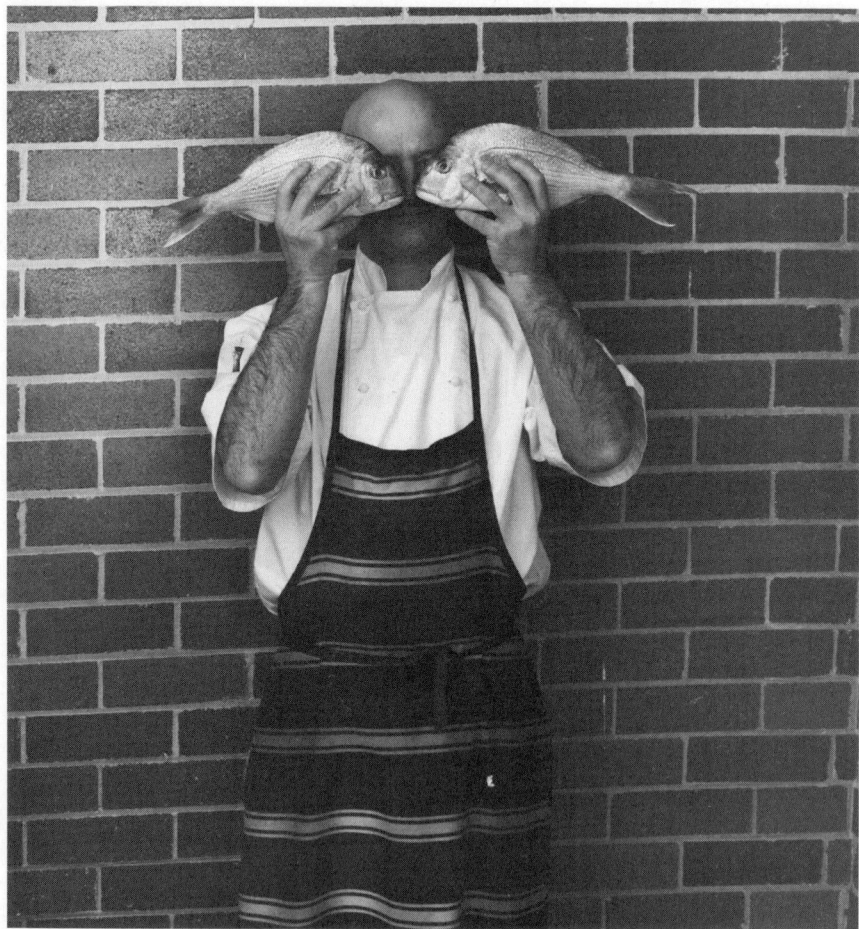

IN the cities and the suburbs of the affluent world, the fish are waiting. Across the cold counters of supermarkets and specialist costermongers, fillets lie translucent on the ice, sparkling like champagne; effete king prawns in pretty pink piles, squid in creamy ringlets, mounds of scallop cushions and whole fish – pan-sized to please and fanning out in geometric formation – give the very impression of a living shoal twisting and dancing above the ocean floor. Octopus tentacles crowd through pots of vinegar and oil; tuna steaks as dark as red wine lie next to white swordfish, mackerel with skin of shining blue tinsel and flatfish patterned as intricately as a Persian rug. Down the aisles are cans of tuna and herring in reassuring rows, sardines playing sardines, sumptuously greasy quilts of smoked salmon in cling wrap; and in the freezers, seafood so coiffured, quiffed and made-over that the balls, nuggets, patties, fingers and sticks are a pageant of taste and form.

Out on the boulevards, the bistros promise the catch of the day, imperial seafood platters and nature's own savoury piñata, the lobster special; while for the budget-minded there is always fish and chips, fillet burgers under the golden arches or a pizza with extra anchovies. Fashion-conscious urbanites find caloric solace and lifestyle choice in sushi packed like coloured candy, or fallen balloons of rice-noodle dumplings that no late morning of dim sum would be complete without. Last thing at night bright amber beads of omega-3 are swallowed down whole. There is even a small oily tin for the cat. It is an uncanny abundance.

'When we look at a piece of fish on our plate,' asks the leading British journalist Charles Clover in lightly censorious tone, 'what do we know about the fish?' Amid the amplitude, the question is apt and disquieting. We indulge magnificently, but more often than not, we know not what we eat. Mostly, we don't know where our seafood has come from. We don't know how it has been caught. We are ignorant about what else may have been damaged along the way.

Clover, for many years the environment editor of the London *Daily Telegraph*, has been preoccupied with documenting what we don't know about fish for around two decades. At first sight, he appears an unlikely marine environmentalist. Devoid of dreadlocks and rainbow-dolphin paraphernalia,

---

PREVIOUS PAGE: *See food (detail).* Photographer: James Dore
from the *Sydney Morning Herald* Shoot the Chef! 2009 photographic competition.

he is instead a mixture of sparkle and starch. Quite the patrician (he once co-authored a book on organic food with Prince Charles), Clover is nonetheless unfailingly phlegmatic. Years of combating 'blue-wash' have left him intolerant of fools and foolishness. He doesn't take bullshit kindly. In 2005, Clover published *The End of the Line: How Overfishing is Changing the World* (Ebury), a seminal work that is lively, accessible and angry, yet seasoned with peppery black humour. In 2009, the book was adapted for screen by Rupert Murray, receiving a widespread release and a slew of positive reviews. Both book and film expose the ugliness and calamity that swim beneath our decadence.

There is a reason why the origin of the seafood on our plates is so often hidden. Under the euphemism of fishing, in pursuit of seafood, we are exterminating animals from the oceans: wiping out the larger predators, altering food webs and destroying habitats at a staggering rate and on a confounding scale, in an epic slaughter of life beneath the waves. The truth taints the palate. Starting in the fresh waters, in rivers and lakes, spreading to the coastal shallows and now rampant in the high and deep seas, we have stalked the stocks to the point of catastrophe. It is a darkly grand story of the reduction of our collective inheritance, and one little apprehended under the glaring lights of the fish-market floor.

WORLDWIDE, STOCKS OF the great fish have been reduced by 90 per cent from 1950s levels. King cod is gone from the Grand Banks. Bluefin tuna is on the brink in the Mediterranean. The great billfish, such as marlin and swordfish, have been radically reduced in numbers. Sharks have become more hunted than hunter. Smaller species are in trouble, too: around two-thirds of all fisheries exploited since the 1950s have collapsed. The overall global catch of all fish has now been in decline for two decades. One reputable, if not uncontroversial, study has forecast the disintegration of all presently exploited commercial fisheries before 2050.

The axiom that what happens at sea stays at sea is apt for more than a cruise romance. It is notoriously difficult to find out what actually goes on in commercial fishing. Occasionally, though, evidence becomes available which brings matters plainly into the open. The *Prolific* is a plucky-looking trawler, painted in bright red and white, like a boat from a children's book. On a calm day in August 2008, the Norwegian coastguard encountered the British-owned craft and decided to film what they saw. The short movie depicts the *Prolific* expelling from its hold a steady stream of good-sized whole fish, dead. Fish dumping – 'discarding' is the jargon – is not allowed in Norwegian waters, so the *Prolific* had simply crossed back over the nautical border. In filming the boat, the Norwegian coastguards were seeking to document the travesty.

The *Prolific* is shown dumping more than five thousand kilograms of freshly caught cod and other white fish into the ocean. The recording is unspectacular and at first does not seem to make sense. A trawler dumping fish: surely it is supposed to be catching them? At the very least, we think, something must have gone wrong: refrigeration failing, for example.

As the film continues, the flow of dead fish is like an absurd joke that runs on too long. Later, the action shifts as the crew begin dumping fish from boxes heavy enough to require two men to lift. While their faces cannot be made out, their body language seems resigned. It seems as if what the crew is doing is not particularly unusual. It is just part of the job.

What we witness in the film is not shocking because it is an aberration, but because it is banal. Every year, somewhere between a quarter and a third of everything caught through fishing worldwide is dumped back over the side, dead or dying. It is hard not to think of the catchy advertising boast: 'It's the fish John West reject that make John West the best.'

Some discarding is perversely mandated by regulations which impose fixed quotas in mixed fisheries, leaving no alternative for fishermen but to consign overboard what they cannot lawfully land. And there is the profit motive of some individual captains who engage in the iniquitous but economically rational practice of 'high-grading', which involves dumping all but the most profitable sizes of the most precious fish in order to maximise their return for each trip to sea. It is no accident that so much of what is sold commercially can fit comfortably into a conventional frying pan. Then there is the collateral damage of our war on fish: the 'by-catch' that nobody wants at all. Depending on the fishery, the refuse pile includes everything from giant turtles to sharks, unmarketable fish including juveniles, starfish, corals and even dolphins or small whales.

The quantity of 'by-catch' is partly a product of the global fishing toolkit. For instance, huge tuna purse seines set under sophisticated fish-aggregation devices scoop up innumerable other animals along with their shoaling target. Some prawn runs are so indiscriminate that more than four-fifths of everything caught is hurled back, maimed or dead, into the seas. Bottom-trawlers are among the most egregiously destructive, sweeping giant nets across the ocean floor, obliterating the seabed and dragging up everything, leaving mud and blood, rubble and fragments in their wake. The demolition has spread to the deep seas, where boats with 10,000-horsepower engines pull nets held open by trawl doors that weigh up to five tonnes. Worldwide, efforts to proscribe the carnage have met with limited success.

17 DECEMBER 2007 was the first day of the annual meeting of European Union fisheries ministers at the Justus Lipsius building, in Brussels. The building, in the heart of the Eurocrat quarter of the Belgian capital, has been the seat of the Council of the European Union for around fifteen years. At about 7 am, activists from fourteen European countries descended on the building, constructing across the main entrance a wall of blocks and cement thirty metres long and more than two metres high. Other entrances were also impeded, preventing free entry to the building. The remarkable effort in construction took forty or so minutes, during which time the builders were largely unhindered by the Belgian constabulary. It seemed fantastic that the work teams had been able to raise the edifice so fast and without interference. An Englishman passing by in a business suit was impressed by the standard of the craft: 'Nice bit of brick work, that,' he observed.

Eventually the police arrived in force to arrest all two hundred activists, many of whom had by then chained themselves together as a human barrier in front of the wall. One by one, the men and women were removed, decoupled from the human chain and dragged off to waiting wagons. Proceedings ended when the last of the arrested had been driven away for processing. The wall remained for a time, emblazoned with the slogan 'Shut down until fish stocks recover', but the demolition gang moved only slightly less swiftly than the construction team, and by midday little trace was left. The action was designed to prevent the EU ministers setting the annual catch quotas for European waters, an exercise in cynical bargaining which decides how much the fishing fleets of the various member states are permitted to catch.

The wall building caused only a short delay in proceedings; but then, the real purpose had been to draw public, media and political attention to the issue in an effort to force a shift in policymaking. Some press interest was briefly focused across the continent, but there was no sudden change in political and policy direction. The great fish trade-off was completed as usual, and catch limits were set well above scientific advice. Over the past two decades the council of European fisheries ministers has approved catch limits exceeding scientific recommendations by an average of 15 to 30 per cent, depending on the species of fish. Like glass, science does not bend and by the EU's own admission, almost nine-tenths of European fish stocks are now over-fished: an environmental, economic and cultural debacle.

ACROSS TIME, THE ocean has been imagined as a foreign dominion, ruled by gods, spirits or fantastic creatures, impervious to terrestrial power. But these myths have vanished in modernity's great unmaking. The failure of the European Union's common fisheries policy to guarantee healthy stocks is symptomatic of a wider global malfunctioning. Like the hosts of a party so committed to accommodating guests as to be too embarrassed to restrain debauchery, fisheries ministers and officials preside over a multi-layered regime of wild excess. No doubt there are good and true men and women among them. Some countries perform better than others. Nonetheless, taken as a whole, there is a dominance of vested interests so complete as to make the sanctioning of the rape of the oceans an abiding political fact.

The world's fishing fleet is two and a half times larger than the oceans can sustainably support. Yet, more than a third of the global fishing industry's revenues come from subsidies, which are currently estimated at around US$34 billion a year. We are funding the fishing industry to systematically destroy our shared inheritance.

The perverse illogic of the global fishing complex can be traced partly to the concentration of fishermen in and around port towns, creating identifiable block votes in specific electorates. Even when the fish are gone, localised voting blocks remain, all the more shrill because the writing is on the wall. As Charles Clover acidly observes, in Britain there is still a fisheries minister despite the wild-capture fishing sector being roughly the size of the lawnmower industry. But looming behind the decreasing numbers of fishermen, and their families, employees and suppliers, are powerful ghosts. Implanted deep within the collective memory is an ideal type, doughty, bearded and gruff, in a sou'wester, the captain of a little boat, plying an honest fishing trade in a struggle with the vicissitudes of the sea. Collective identities are tied up in the myth of the noble fisherman, whose bravery and endurance commands the respect of unadventurous and weak-kneed landlubbers.

Small-scale fishermen still exist, but the industrial might of the global fishing industry bears little resemblance to old man Santiago. Much of the planet's seafood is caught and bought by transnational corporate giants, as blind to the long-term interests of small-scale fishermen as the wind and waves, and just as amoral. It is sometimes said that fishermen are natural conservationists, because it cannot be in their long-term interests if fish stocks collapse, a claim which undoubtedly has some truth in local and traditional fishing communities. Overwhelmingly, though, history suggests that once capital is in play, wolves make poor guards of sheep. In an age of fluid global finance, the exhaustion of a fish stock means no more to a corporate investor than an indication that it is time to seek a new venture for increasing return. Whatever normative restraints that apply to a traditional fishing community are inapplicable to a mega-trawler flagged to a nation on the other side of the world and owned by a corporation, a non-human abstraction that is mandated to do no more than maximise profit.

THE RETAIL DISPLAYS of seafood shout of plenitude, but there are spectres at our feast. The oceans are not exempted from the political economy of global resource distribution. No matter the overall scarcity, the developed world's bowl is always full. Japan is the world's single largest seafood importer, eating far more fish than remain within its territorial waters. Australia's fisheries are among the better managed, but the country is still a net importer of seafood. Europe's shortfall has to be made up from somewhere. In the coastal waters of developing countries, trawlers sweep through, retaining the most lucrative fish for first-world tables, but smashing and killing much else in their path, depriving developing populations of critical food resources. Stocks off the west coast of Africa are thought to have halved since 1945.

Clover saves special rancour for the wastefulness of aquaculture. Intuitively, fish farming suggests a way of reducing pressure on the wild resource. In practice, the cultivation of fish often makes matters worse, adding new layers of exploitation. In order to feed the salmon, prawns and other farmed seafood, feed is required, which entails catching vast quantities of smaller and less marketable animals from the ocean, then reducing them to meal. In what Clover describes as a 'new circle of hell', edible fish are diverted from the mouths of the developing world to the fish farms of the north.

Beneath the gilded profusion of seafood for our dining pleasure lies a deeper irony: for all the ubiquity and convenience, our menu is but a pale imitation of the riches enjoyed by previous generations. One of the scientists featured in the film *The End of the Line* is the softly spoken and smiley Professor Callum Roberts. In his work, Roberts has examined the scientific, archaeological and historical record to establish changes in patterns of seafood capture over time. Because of our 'collective amnesia', writes Roberts in his magisterial and essential *The Unnatural History of the Sea* (Gaia Books, 2007), 'few people really appreciate how far the oceans have been altered from their pre-exploitation state.' He concludes that stories handed down, of enormous fish and vast shoals writhing, unthinkable today, were not the tales of addled sailors but sober descriptions of actual experience. We have, in the language of marine biologists, fished down the food chain, reducing or eliminating layers of predators in turn, but the transformation is barely noticed as 'new species' take the place of those no longer available. As Professor Daniel Pauly of the University of British Columbia has put it, 'We are eating today what our grandparents used as bait.'

EYES WIDE OPEN, dead fish always look surprised. Destructive fishing is the most serious environmental threat to the oceans that comes from our activities on the water, but it is compounded by other dangers. The spread of pollution in the oceans is both ubiquitous and concentrated. In the North Pacific, the trash gyre of plastic and other rubbish has now grown to cover an area of ocean about three times the size of Victoria. Then there is climate change. The impact of global warming on the ocean is potentially severe: raising water temperatures, disrupting currents, melting ice and altering chemical composition to increase acidity. We are in the realm of diabolical consequences and unpredictable feedback loops; it simply is not clear how weird things might get if climate change cannot be brought under control. Meanwhile, the depravity of the plunder is unabated. In *The End of the Line*, Clover documents an instance in North America of fish being rendered into oil that is burned as fuel. Recently, technology has been pioneered that allows the tiny krill, minute shrimps at the base of the food chain in the Southern Ocean, to be vacuumed from the water. In darker moments, scientists warn of a return to primordial oceans, dominated by invertebrates. We face the long night of an aquatic dystopia of our own making.

If the world's fish stocks are to be renewed, we must create great marine reserves, spanning the world's oceans. Like national parks at sea, marine reserves can let ecosystems recover, generating to a new profusion of life. The campaign must be monumental: in forty scientific studies examining how much of the world's oceans should be protected, the majority indicated a range between 20 and 50 per cent, but currently the figure languishes at less than 1 per cent.

In those instances where marine reserves have been established, offering full protection from fishing, the increase in fish stocks is remarkable, growing five and tenfold in a decade or less. Fishermen have been shown to benefit too, because marine reserves act as nodes of fish production, restocking the surrounding waters. Other measures are also needed: fleet reductions; an end to perverse subsidies and disinterested adherence to scientific advice on catch limits; greater restrictions and improved enforcement; the banning of certain gear types; and improved certification requirements.

As consumers we can and should choose more sustainable seafood, but we can't simply buy our way out of trouble. Ethical purchasing alone is insufficient. Ultimately, if we are to redeem our oceans, we will have to engage in a political contest. Vested interests and timid tolerance of the status quo must be confronted, in order to secure the common wealth. Beneath the surface and sinking, with hope and will we can still strike back upwards toward the light.

References available at www.griffithreview.com

David Ritter is a visiting fellow in the Faculty of Law at the University of Western Australia and the head of biodiversity at Greenpeace in London. He was formerly a leading Australian Indigenous-rights lawyer, and his most recent book is *Contesting Native Title* (Allen & Unwin, 2009).

How many miles?

Out of Africa

Tony Barrell

AT the delicatessen counter of my local Woolworths supermarket – which promotes itself as 'the fresh food people' – in the inner-Sydney suburb of Balmain, I saw some fillets of firm white-fleshed fish for sale. They were, said the caption on the tray, 'Nile perch' imported 'frozen' from Uganda. I found this hard to believe, but the counter-hand confirmed it. Yes, they were 'fully imported'. Well, at least Woolworths says where its food comes from.

A week later I visited a small fish shop in the same suburb that sells a few fillets and shellfish. I asked for a handful of scallops, not noticing the caption. The man at the counter told me with some pride that they were 'Japanese'.

'Japanese scallops?' I said. 'We are importing scallops from Japan?'

'Yes,' he said.

'Why?' I asked.

'Because they're cheaper,' said he.

'Cheaper? How can that be?' Caught in Japan's inland sea, frozen and then flown here, were they really still cheaper than ours, from Tasmania? That week our dollar was buying only ¥60! (It's improved a lot since.)

'They're probably farmed,' said the counter-man.

'No doubt,' I replied, and wondered whether they were farmed in Japan or imported there from Japanese-owned scallop ponds somewhere in South-East Asia. I left with a handful of prawns, not frozen but probably farmed.

I later discovered – through David Hardy's *Scallop Farming* (Blackwell, 2006) – that scallops in Japan have traditionally been grown and fattened by aqua-cultural co-ops around northern shorelines like Mutsu Bay, on the main island of Honshū. They are attached to long-line nets or baskets that trail down for two hundred metres or so, and don't need to be fed or protected with chemicals. Within the aqua-cultural community the Japanese scallop farm is regarded as a model of efficiency. So there.

Nevertheless, the implications of food miles are baffling. I once filmed the harvesting and snap-freezing of prawns straight from brackish ponds in Luzon, in the Philippines, and knew that Japanese seafood importers bought direct from these farms throughout South-East Asia. Conglomerates such as Maruha – formerly Taiyo, and involved in whale hunting – import seafood from all over the world. (Taiyo used to own a baseball team, the Taiyo Whales, but the team changed its name to the BayStars when the anti-whaling movement started in the US.)

I investigated what else was on offer in my part of Sydney. At a bigger Woolworths, in Leichhardt Market Place, I found more Nile perch fillets, priced at $11.98 ('from Lake Victoria imp frzn produce of Uganda/Tanzania/Kenya'). There was also 'basa' from Vietnam, prawns from Thailand and China, barramundi from Taiwan and smoked cod from South Africa – all of which would have arrived here frozen, even though the permanent signage above the section says 'Fresh Seafood'. In the same centre, Aldi was offering a variety of frozen 'white fish' in packets, imported from New Zealand. Closer examination revealed these to be 'red cod' (skinless or with 'skin on') and 'hoki' (blue grenadier). At Coles, smoked cod fillets ('capensis hake', 'wild caught' and 'thawed') were also available, at $11.98.

Jonnie's, a smaller specialist shop, also offers Vietnamese 'basa', and I have since seen the fish in several fish markets. The man behind the counter said that it was imported frozen and defrosted by Jonnie. At Sydney's fish markets, as well as the wide range of fresh fish caught in Australian and New Zealand waters, you can also buy dried and frozen fish, including smoked cod.

As for the Nile perch: they don't come from the great river but from Lake Victoria, into which they were probably introduced, and are caught by fishermen from Kenya and Tanzania – creating their main export earnings. Nile perch is also known as black bass – the species name is *Lates niloticus* – and can weigh more than two hundred kilograms. You can go game fishing for the 'world's largest freshwater fish' near the Ssese Islands, in the centre of the lake. And you can buy fillets online, direct from Uganda, for about $5.50 a kilo – if you order a minimum of eighteen thousand kilograms. Woolworths doesn't buy from this supplier, and its prices have ranged this year from $11.98 to $16.98 – cheap, when you consider that fresh local flathead fillets cost more than thirty dollars.

Uganda's Nile perch trade expanded in the 1990s and doubled in recent times. It now constitutes 90 per cent of the country's fish exports, most of which go to Europe. The Portuguese fisheries consultancy Megapesca reports that because Europeans, Americans and Australians like fish without much fat, the perch are 'deep-skinned' the day after landing, when all 'dark flesh' is also removed. The Japanese prefer theirs with the skin still on. The perch are then 'blast-frozen' and trucked to the port of Mombasa, or airports at Entebbe or Nairobi, from which they are flown to Europe and elsewhere at a cost of up to US$2 a kilogram.

Uganda also exports a smaller fish to Europe, the more plentiful tilapia, also frozen; but despite a short-term moratorium on exports to Europe ten years ago, when it was first realised Lake Victoria was being over-fished, Nile perch stocks are again being seriously depleted. According to Fish & Information Services Australia they fell by more than 40 per cent in 2008; ten fish processing factories have already closed, and the Lake Victoria Fisheries Organization has been asked to impose either a seasonal moratorium on fishing part of the lake or a total ban for two years, and a crackdown on poachers and the expansion of fish farms.

I confess to feeling squeamish about eating the carnivorous *Lates niloticus*, even if it is rich in the omega-3 fatty acids we all need, because I wonder how old it might be and what it might have eaten. But clearly there are Australian customers who buy it.

I wrote to Woolworths and asked them why they still imported Nile perch from East Africa. The answers I got were, to say the least, opaque. Yes, they were aware that the fish was in short supply – although there was no acknowledgement that as a species it was 'threatened' – but that was due to the European trade, and in response Woolworths is no longer promoting the product. In the words of its media department: 'As much of the Nile perch stock is heading to Europe we have therefore decided to stop advertising Nile perch to reduce demand for it through our stores. But while there is still a consumer demand for it we will continue to stock it.'

I couldn't tell whether this meant Woolworths was running down stock and stopping imports, so I asked again. 'Forgive me for seeming obtuse but is Woolworths running down existing stocks or still ordering supplies from Africa? Hope the question is clear.'

But answer came there none. Nile perch fillets are still on display, and were $16.49 when last spotted.

Tony Barrell is a writer and broadcaster. His work has appeared in *Griffith REVIEW: Up North*, *Cities on the Edge*, *MoneySexPower* and *After the Crisis*.

For more about the Nile perch see www.aquaticcommunity.com/mix/nileperch.php

Trouble at dolphin cove

Eating whale in Japan

Cory Taylor

IN a story by Ryunosuke Akutagawa called 'The Spider Thread', the thief and arsonist Kandata is writhing in hell with all the other sinners when the Lord Buddha happens to look down from paradise. Buddha knows, as one who knows everything misses nothing, that Kandata once saved the life of a spider on a mountain trail. He took pity on a spider that it was in his power to squash underfoot, figuring that even the spider possessed a life worthy of some respect.

On the strength of this single instant of compassion in a lifetime of vileness, Buddha is prepared to throw Kandata a lifeline. He takes a spider thread and sends it floating down on the heavenly breeze in his direction. A skilled climber, the career housebreaker grabs hold of the thread and starts hauling himself up out of the hellish depths, full of glee at this surprising change of fortune.

And then he makes a fatal error. He looks down. The delicate thread is supporting not just him but thousands of other wretches, sinners desperately scrambling over each other to escape the terrible torments of hell. Acting quickly, Kandata snaps the spider thread below him and sends them all back down into a river of blood. The Buddha, according to Akutagawa, is a little saddened.

I am not a Buddhist, but I think we are meant to infer from this divine melancholy that the Lord Buddha had hopes of a better outcome, some evidence that Kandata's compassion towards the spider spoke of a larger capacity for compassion towards his fellow sufferers in hell. Now, however, there is nothing for the all-knowing to do but to snap the spider thread and send the thief tumbling back down.

The moral of the story is at least twofold. Even the smallest act of compassion is looked upon with favour, just as any act that is devoid of compassion ensures that one's suffering will continue unrelieved forever.

THERE IS SOMETHING of the thief in Ric O'Barry, star of a recent documentary, *The Cove*, which had its first Japanese outing at the Tokyo International Film Festival in October 2009, two months after its cinema release in Australia. I am old enough to remember O'Barry in his first incarnation, as Flipper's trainer in the television series of the same name. For ten years, O'Barry lived the high life as a result of the dolphin's popularity, until something happened to Flipper, or at least to one of the dolphins that played the lead, the one he called Kathy.

O'Barry had known this dolphin intimately. He had captured her, trained her, watched her give birth, nursed her back to health when she was sick. Then Kathy died in O'Barry's arms – a suicide, according to his commentary in the film. It was provoked, O'Barry believes, by the stress and depression that all dolphins suffer, and ultimately die from, when kept in captivity.

Kathy's demise was the turning point in O'Barry's life. Since her death, he has made it his mission to shut down the vast and lucrative trade in live dolphins that *Flipper* spawned. He does this by any means possible: by agitating at international forums on whaling, by rescuing captive dolphins, by blowing the lid on dolphin hunting using all the high-tech, gee-whiz surveillance camera and sound gear available to him. His compassion for dolphins is a formidable weapon that he wields with all the zeal of the convert, a man who once was lost but now is found.

Taiji, a small fishing village on the east coast of Wakayama Prefecture, two and a half hours south of Nagoya, is the setting for O'Barry's latest skirmish with his enemy. Nestled snugly in a remote corner of one of Japan's most scenic coastlines, it boasts a long seafaring history. When whales were still hunted there, it must have been a proud place full of proud men. Up on the wild headland that overlooks the Pacific on one side and the town on the other is a stone pulpit, where a lookout was posted to watch the horizon day and night for the whales' approach. This was a position open only to samurai. No ordinary folk need apply.

Now that the whales are hunted elsewhere, Taiji has turned to dolphins to save its fading fortunes. The town is a world leader in the capture and killing of dolphins, a practice sanctioned and overseen by the Ministry of Agriculture, Forestry and Fisheries in Tokyo, and one that sees dolphins sold to marine parks all over the planet.

The Lord Buddha is no doubt watching events in Taiji with the same kind of sad aloofness he displayed when witnessing Kandata's fate. He is probably, like a lot of others, hoping and despairing in equal measure. I imagine he wishes he could look away, but knows that for him ignorance is not an option.

THERE IS NO nice way of describing what happens to dolphins at Taiji. Akutagawa's river of blood comes to mind. The fishermen hunt the dolphins down by banging on long metal staves, setting up an underwater wall of sound that confuses the dolphins' sensitive acoustic equipment. Panic ensues. The dolphins are harried into a narrow inlet in their scores, then a net is strung between the two shores and the prisoners remain there until the dolphin-buyers arrive to select animals for shipment to amusement parks as far away as Florida and Dubai. Each dolphin will fetch around a hundred and fifty thousand Australian dollars.

The ones that are left are murdered according to a quota system worked out by officials in Tokyo, and fetch six hundred dollars a head. Since this practice has started to attract adverse publicity for the town, the killing is now done in the dead of night, in a hidden cove off the main inlet. It isn't hard to imagine the scene, and thanks to the makers of *The Cove* there is now no need. The men are in their tinnies armed with spears; the animals are in the shallow water at their feet. After a while the water turns scarlet. When all of the animals are dead, their corpses are hauled aboard the boats to be taken back to town, where they are butchered and sold for consumption, although who actually eats the dolphins is hard to verify.

According to O'Barry, the meat is often labelled as whale and sold to unsuspecting whale-meat lovers in local supermarkets, especially in Kyushu, where I live for part of every year. Like a lot of claims made in the film, this one is hard to substantiate and to do so would really be missing the point as, to an audience presumed to be predominantly non-Japanese, there probably isn't a clear moral division between the options anyway. In the minds of most foreigners, eating whale is as damnable as eating dolphin, and in this sense the film's message is uncomplicated to anyone who isn't partial to either.

For a lot of Japanese, the issue of what is edible and what is not cannot be described as a moral one. For Japanese viewers, *The Cove* is much more likely to have raised questions of national pride and cultural heritage than it is to have pricked the collective conscience – particularly given the can-do cowboy tactics the filmmakers spend half the film celebrating.

As they flaunt their questionable methods, the filmmakers reclaim the moral high ground by focusing attention on O'Barry's claim that eating dolphin is dangerous, that the Japanese are inviting a second Minamata if they keep foisting dolphin meat on an unsuspecting public. Footage of the victims of mercury poisoning at Minamata is included to drive home the point. Claims are made that dolphins caught and eaten by the locals are extremely high in toxins, including mercury. Hair samples are taken from a local fisheries spokesman and tested. The results are worrying. The issue is made to seem less about culture and more about public safety, and O'Barry can claim a victory on this score at least. Since agitation on the issue of mercury poisoning began, dolphin meat has been removed from the lunch menu in elementary schools. Compassion for dolphins morphed, by means of mercury, into compassion for Taiji's children.

THIS STRATEGY IS not likely to deflect attention from the cultural issues at the heart of the matter of what is eaten by whom. The question of culture remains, huge and intractable – the elephant whale in the corner. Taiji, as O'Barry says, is probably the only place on earth where you can have your whale and eat it too. Across the road from the whale museum and the marine park, you can buy a whale lunchbox and eat it while you watch the captive dolphins and killer whales go through their paces. Or you can take your lunchbox and wander around to the cove where the slaughter of the dolphins takes place, and sit on the seawall and watch the captive dolphins in the smaller secret inlet circle around in slow motion, trying and failing to make sense of things. Dolphins are smart, but they aren't smart enough. Without proper training of the kind O'Barry popularised, it doesn't occur to them to leap the fence and swim away.

Our lunchboxes were bought at the station in Wakayama, two and a half hours away, and featured not whale but a special style of vinegared sushi rice wrapped in fragrant leaves, and mackerel wrapped in kelp. But we couldn't taste any of it. The cove where we sat is Japan at its most picturesque. We had the place to ourselves. The autumn sky was cloudless. We should have been having fun, but down in the opalesque water a hundred metres from where we sat, there were five fully grown Flippers and one baby Flipper whose fates we knew, and that was enough to rob the day of any joy. Taiji was ruined for us. We already knew more about the place than we wanted to. We revised our plans for the afternoon and decided to get out of town on the next boat, and head for the hills.

The tourist in Japan is special, driven by appetites non-Japanese don't generally comprehend. The Japanese tend not to go to places where there is nothing delectable to eat when they get there. You come to Taiji to eat whale, the same way you go to Hakata, in Kyushu, to eat spicy marinated pollock roe – the local *meibutsu*, or famed delicacy. You come, eat, buy ten packs to take home. If you can't take it yourself, you ask the shop to send it to friends and family by courier halfway across the country, with your compliments. The cult of whale in Taiji has nothing to do with actual whaling anymore. It is about the importance of eating in Japan, and the lifeline this provides for local micro-economies all over the country.

IN LATE OCTOBER 2009, the first section of the new main train station in Hakata, a ten-minute walk from my place, was opened. There are no new train lines, just a development that will see the current building – already a labyrinthine burrow lined with eateries and souvenir shops – massively extended so that more can be added. The section that has just been opened is divided into Gourmet Street on Level B2, the second basement down, Souvenir Street on Level B1 and Drink Street on the ground floor. When we went there the place was packed. Tiny bars full of businessmen with trains to catch were swilling the famous local *shochu* and spooning down lashings of offal stew, which is as famous in Hakata as spicy marinated pollock roe. Whale was on the menu too, as it is in every corner bar, but in our town it lacks the star billing that, in Taiji, makes it seem to be the only thing on the menu.

The queue outside the crepe shop on Gourmet Street was a mile long. This was what the girls in our town had come to savour: Japanese-style French crepes a dollar each while they lasted. In the morning the news must have gone out with the daily papers, on one of those gaudy printed sheets full of food specials: sushi day at the local discount supermarket, three platters for a thousand yen, pork-cutlet dinner for ¥390 until the end of the month, free curry and rice if you can find and present your ANA boarding pass.

My Japanese husband keeps a magnifying glass on the kitchen table so he can read the maps – a lot of these places are hard to find, but worth the search. It is difficult to eat badly in Japan. You would have to ignore every sign and resist every invocation to taste this or that dish of whatever it is that you have never seen prepared in just this way before.

It would be easy to dismiss the cult of food here as merely the supreme expression of a bloated advertising and media industry. Japan, my husband tells me, is a *kajyo* society, best translated as 'glut'. There is too much of everything and too much information about all of it. Food is no exception.

I don't know what the Lord Buddha makes of all this over-consumption. He is usually depicted as well fed, even fat, but at the same time serene. He doesn't look like the kind of deity to harbour puritan thoughts on food, which possibly explains his success in so many cultures where scarcity is the norm.

THIS BRINGS US to the reason the Japanese hunt and eat whale. Even if the official line is that whaling is for science, this doesn't explain how so much whale sashimi, or dolphin in disguise, finds its way onto so many supermarket trays and so many hole-in-the-wall *izakaya* kitchens. The numbers differ, depending on who is counting. Opponents of whaling argue that recent rises in Japanese stockpiles of frozen whale meat reflect a fall in demand. Supporters of whaling argue that rising stockpiles reflect changes to whaling quotas and that demand is growing in line with supply. Even so, the figures are unspectacular. According to Nippon Research Center figures for 2006, some 95 per cent of Japanese have never eaten whale, or have eaten it only rarely. The numbers are dismal enough to have prompted occasional handouts of frozen stocks to schools, in the hope that a new generation might be encouraged to go for whale meat in a big way.

Taiji's fierce attachment to the whale, as both its town emblem and favourite food, seems misplaced if not perverse, as do rumours that the town's fishermen are planning to extend and improve the facilities for butchering and marketing dolphin. Whales were probably first eaten off the beaches in ancient Taiji because there was very little else. Japan is only recently a rich country, and is now deeply divided between the rich cities and the poor and ageing countryside. Anybody over sixty-five — and a disproportionate number of Japanese are — remembers when starvation was a real and visible threat to life. Even fishermen must have struggled. At the war's end the sea was still full of fish, even if the towns had been bombed out of existence or burned to the ground, but without fishing boats there was no way to catch them. The boats and the men who sailed them had all been drafted for navy service. Many never came back.

There is no way to reassure such people that these times will never return. Food security is a potent national anxiety here, given that one of the consequences of Japan's postwar prosperity has been increasing dependency on imported food. The defence of whaling stems in part from the perception that Japan must maintain its right to catch and eat whatever it can, in case there comes a time when foreign supplies, for whatever reason, start to dry up. The supreme irony of the current troubles in Taiji is that America is partly to blame, as it was America that encouraged a revival in the Japanese whaling industry after the war, and in effect popularised whale as a food source to a people on the brink of famine.

Kumi Kato teaches Tourism at Wakayama University. She is interested in sustainability and in spirituality as it relates to the environment. She invited us to a meeting with some of the local Taiji elders to discuss how best to respond to *The Cove*, given that in the eyes of the outside world their town is a gateway to dolphin hell. To them it felt as if a thief had entered their house and stomped around breaking things, before leaving empty-handed. The fact is, Kumi Kato tells the elders as gently as possible, they are going to have to find a better way to represent themselves to the world, a means to avoid the wily defensiveness the makers of *The Cove* have constructed for them as a default position. For Kato, the issue is whether the voices of Taiji's people have any chance of being heard above the din created by the film.

As a small first step, she suggests they invite local and international artists to come and live among them, to create positive relationships that take time and effort, rather than fly-in-fly-out confrontations like those in which activists specialise. Everybody agrees that this is a good idea, that cultural sensitivities have been badly damaged and will need help to repair.

We are asked what Australians make of Broome's reaction to the film, which was initially angry and is now conciliatory, in a confused kind of way. On the table in front of us are the minutes from the last meeting of the Broome Town Council. They include an unqualified apology from the council to the citizens of Broome with Japanese ancestry for any offence caused to them by remarks made at a previous meeting on the subject of *The Cove*. Whatever was said may, it is noted, have provoked racially motivated attacks on the town's Japanese cemetery. And for this the councillors are sincerely sorry.

The minutes also record a vote to immediately undo the ill-considered ending of the sister-city relationship between the two cities, while at the same time condemning the killing of dolphins. The gesture of conciliation is warmly appreciated, the condemnation of the dolphin hunt passed over in silence.

There are links between Taiji and far-flung towns like Broome all over the world, testament to generations of Taiji men who left to seek their fortunes as whalers, fishermen, sailors and divers for pearl shell anywhere there were prospects for betterment. One Taiji elder has just come back from San Pedro, California, where he met some of their descendants. There are many more scattered all over South America, Australia and the South Pacific.

We tell our hosts we are just as confused as they are, that Broome is a long way from anywhere and that the council's pronouncements are not an expression of a national consensus.

It is hard to know who is more deserving of compassion in this messy tale, or whose cross is the heaviest to bear. The makers of *The Cove* have no doubt that the dolphins deserve to be pitied the most, as they are so clearly defenceless and so handsome, nature at its most charismatic. The fishermen of Taiji are apparently unmoved. Dolphins, as far as they are concerned, are a livelihood for the whole town. Without the dolphin hunt, Taiji would shut down.

It is precariously close to doing so anyway. There are shuttered businesses and abandoned houses wherever you look, just as there are in all the towns you pass along the winding train line to get there. Japan's prolonged recession is taking its toll, nowhere more so than in backwoods towns with little to attract the scarce tourist dollar now that travel to other countries in Asia and elsewhere is so much cheaper. Instead of travelling to Taiji for the weekend, we could have gone to Seoul for half the price – around A\$350, including airfares and a succulent Korean barbecue.

For his part, Ric O'Barry is trapped in a purgatory of his own making, forever the mastermind behind an entertainment juggernaut that exploited captive creatures he now believes have souls as evolved as our own.

Buddha has probably mulled over paradoxes like his on countless occasions. It is hard to know how he makes up his mind in cases so morally complex as this – how Buddha decides who should be condemned for eternity and who deserves a second chance. If Akutagawa's story is any guide, this deity's methods are as mysterious as those of any other. Having sent the thief Kandata down to hell with all the other poor sinners, The Lord Buddha, it is written, continued on his way down the divine pathways of paradise with a faint smile on his face. 'I think,' the story closes, 'it was about midday in paradise.'

It was lunchtime in Taiji when the meeting with the local elders ended. We could tell because there was a notice posted on the entrance to the Citizen's Hall as we left the building, inviting the townspeople to try the special whale lunchbox *available today only between twelve and two* at the fisherman's co-op down the street. A chance, the notice said, not to be missed.

Cory Taylor is a freelance writer based in Brisbane and Hakata, Japan. She has a PhD in film and television from Queensland University of Technology and teaches art theory at Queensland College of Art.

'A wild, disquieting ride through
the back roads of the new China.'
NICOLAS ROTHWELL

'Hessler bares the country's heart and soul with
grace, humour and rare modesty.'
ROBERT MACKLIN

'In his latest feat of penetrating social reportage,
New Yorker writer Hessler again proves himself
America's keenest observer of the new China.'
PW REVIEW

PETER HESSLER

COUNTRY DRIVING

THREE JOURNEYS ACROSS
A CHANGING CHINA

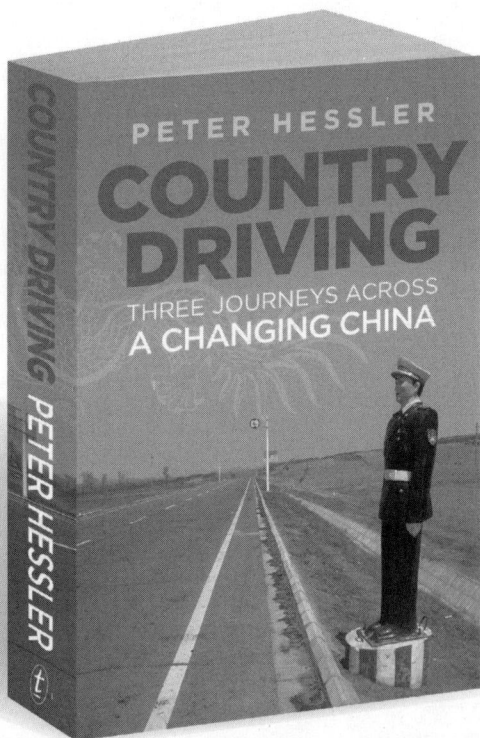

Text Publishing, Melbourne Australia
WWW.TEXTPUBLISHING.COM.AU

Re-thinking animals

The next rights frontier

Anne Coombs

'ALEX taught me to believe that his little bird brain was conscious in some manner; that is, capable of intention. By extrapolation, Alex taught me that we live in a world populated by thinking, conscious creatures. Not human thinking. Not human consciousness. But not mindless automatons sleepwalking through their lives, either.' So Irene Pepperberg wrote about her African grey parrot Alex, the most famous parrot in history. Alex died in 2007, after working with Pepperberg in her laboratory for thirty years. He was already well known, but when he died he got obituaries in the *New York Times* and *Time*.

When Pepperberg began working with Alex, she had no idea that he was going to help overturn human understanding of the capabilities of animals. Alex showed what fond pet owners have long suspected: there is a lot going on inside an animal, and we can only guess the half of it. For hundreds of years, animals have been viewed as less important, less feeling, less cognisant. For most of the twentieth century, scientists derided the notion of animal consciousness: dumb animals – not able to speak, not able to think. Language was seen as the magic attribute that separated humans from brutes. But over the past thirty or forty years, scientists who have been working with non-human animals have been gradually uncovering the special talents and means of communications of all sorts of animals, from bees to octopi, dolphins to elephants.

Recognising that animals have consciousness is a huge step for humans, and dangerous, too, as it might well undermine not only traditional thinking but our traditional means of survival – that is, what we eat.

For much of humankind's history, animal suffering didn't matter. Some religions give animals consideration in theory, but there are still many places in the world where animal suffering, even torture, is commonplace. Yet the position of animals, their place in the universe and our responsibility towards them, has been a perpetual question for philosophers. Mostly their conclusions have not been to animals' advantage. The idea of the food chain, a hierarchy of value, originated with St Thomas Aquinas and his belief in animals as mere instruments of humans. Immanuel Kant said that because animals lacked autonomy and rationality, they had no moral value. And Rene Descartes denied animal consciousness on the basis that it was not necessary to believe in animals' awareness in order to explain their behaviour. Human belief that animals don't suffer – or at least not as we suffer – has justified much cruelty.

IN JULY 2009, several hundred delegates – philosophers, scientists, animal-rights activists, historians – from more than twenty countries descended on Newcastle, in New South Wales, for the largest animal-studies conference ever held. 'Minding Animals' was held at the City Hall, a wonderful 1930s edifice that seems to exude confidence about human achievements and satis-faction with our place in the world as the pre-eminent species. A startling array of papers was delivered in five or six simultaneous sessions every hour for six days.

There is no one academic area that covers 'animal studies'. It is a sprawl-ing subject that spills over from ethology (animal behaviour) into biology, psychology, geography, neuro- and cognitive science, and veterinary science. And then there are the humanities – history, literature, religion, philosophy – where animals crop up constantly, as soon as you start to look. There is so much going on in animal studies that it seems inevitable it will change the way that humans view animals. Already it is beginning to affect what we eat and how we farm animals.

Philosophers are at the forefront of much of the work. Peter Singer, whose 1975 book *Animal Liberation* (Random House) became a founding text of the animal-rights movement, told the conference that over the past three decades philosophers have challenged people to re-examine their beliefs about animals; in particular, to recognise the 'cognitive overlap' between humans and animals. On the matter of eating them, he said it was unrealistic to think that everyone would become a vegetarian, but it was possible for everyone to eat less meat, and that in itself would reduce animal suffering.

The food at the conference was vegan, and this was popular even among the meat-eaters. It is not possible to sit for hours hearing about the lives of animals, then turn around and eat one.

I took my first steps towards vegetarianism at that conference. I don't think I will ever become a strict vegetarian, but I now think about what I eat, and once you start to do that you find that there are a lot of dishes that you simply have no appetite for: chickens, for example, the ones that have been grown fat and fast, six weeks from birth to death, in vast broiler sheds, their little skeletons unable to support the artificial growth forced upon them. Or pork which comes from an industry where sows are piglet machines and spend their lives in cages so small they cannot turn around in them.

It is now widely accepted that animals do feel pain, but the degree of their ability to feel, know or perceive – their sentience and cognisance – is still contested. Apes and dolphins, maybe, but lizards? We know that the individuals of many species recognise one another, communicate with one another, feel distress and sorrow. Maybe even lizards, although the jury is still out.

At a session entitled 'Attending to Animals', the point was made that feminism has had little to say about animals, perhaps because as soon as one starts 'attending to animals', concepts like care, nurturing, attentiveness and consideration arise – concepts that in the early years of feminism women were keen to separate from beliefs about what it was to be a woman. But times have changed. Instead of worrying about our relationships with men, some of us are beginning to worry about our relationships with animals. Being attentive to them, and recognising the relationships we have with them, unveils a new world view – the start, perhaps, of another revolution. The first step is recognising that each animal is the subject of its own life.

HUMANS SLEEP IN order to restore our brains after the demands of a day of conscious life. Frogs and other animals that probably aren't conscious don't seem to sleep. But mammals and birds sleep and maybe it is because they, too, need a break from their conscious lives. And maybe they also wake, as we do, and 'experience their presence in the world'.

Alex the parrot was able to prove that he could think and feel. On one occasion, Irene Pepperberg was demonstrating to a visiting colleague some of the things of which Alex was capable: counting and distinguishing between objects. It had become clear to her that Alex sometimes got bored (as well he might, given that he sometimes had to repeat an experiment hundreds of times for it to be scientific). On this particular day, Alex was not being co-operative. He was shown a range of objects in groups – three of this, four of that – and was asked a question, the answer to which was 'two'. But he repeatedly gave every answer except 'two'. So Pepperberg told him, 'Okay, you're going back in your room.' She put him in his room and closed the door. As she walked away, Alex began calling out, 'Two…two…I'm sorry. Come back!'

There are so many mysteries in the minds and bodies of non-human animals, and we are only beginning to get an inkling of them. The more researchers open up their own, and our, thinking about the nature of animals, the more the mysteries deepen.

Petra Stapp, a young English researcher looking into the nature of human-animal encounters, was told of an encounter between a woman and a sheep on a moor. The woman was out walking with her boyfriend. She decided to sit for a while to enjoy the surroundings while her boyfriend walked on. Soon a single sheep appeared over a hill. It walked towards her. Indeed, it walked right up to her, purposefully. She watched. When it reached her, the sheep leaned forward and kissed her on the lips. Astonished, the woman sat still. Sheep and woman observed each other, face to face, for a few moments. Then the sheep turned and retraced its steps.

The boyfriend observed the interaction from further along the track. The two then finished their walk and returned to the car. The woman switched on her mobile phone and retrieved a message left at about the same time as her encounter with the sheep. The message was from a friend, telling her that another close friend had just died.

How do we interpret this? The woman was unlikely to be hallucinating, because the act was observed by a third party. Maybe it was a coincidental meeting with an affectionate sheep. Or perhaps the sheep was, as the woman believed, delivering a farewell from her friend. We have no rational way of accounting for that. We may never know the truth behind such human–animal occurrences. But stories such as this suggest that our human-centric explanations of animal capability are seriously lacking.

OUR LIVES ARE entwined with those of animals. For thousands of years, they have helped us survive in all sorts of ways: dogs who became domesticated and hunted with humans, benefiting both; cats that enabled people to begin accumulating grain because they controlled rats and mice. (Without the ability to store grain, cities might never have developed.) The 'beasts of burden' that made it possible for human societies to move, conquer and colonise.

There is a little-known example of the latter in Australia's history. Camels have been romanticised in the Australian story – 'ships of the desert' – but alongside them, and outnumbering them, were donkeys. Big teams of fifty or more were used to pull the wagons that took supplies and mail. They moved between towns, stations and settlements. When motorised transport came and the animals weren't needed anymore, the men who used and cared for them took them out into the bush to some remote well and let them go. The donkeys, being donkeys – smart and adaptable – not only survived but multiplied.

Jill Bough, a Newcastle historian, has studied what became of them, and what happens to them now. They're what we like to call 'feral', meaning that, having introduced them, we damn them for existing in places adjacent to those we introduced them to. Chased and hunted off good land, they retreated ever further into the interior. They give humans a wide berth.

Aerial culling is widely accepted: the removal of species that are damaging our environment, threatening the livelihood of native animals and the like. But the closer you look, the more complicated it becomes. Once again humans are playing god, deciding which are worthy animals and which are not. It may be that donkeys eat the grass that native animals might eat, but the real problem is that they eat the grass that the cattle belonging to pastoralists might otherwise eat.

Since the 1970s, hundreds of thousands of donkeys in the Kimberley and the Northern Territory have been culled from the air. Since 1994, the culling has been particularly nasty. A female, a mature jenny, is captured and fitted with a radio collar, then released. She meets up with a herd and the helicopter shooters track the herd by following the radio signal. All the donkeys are then shot, bar the jenny with the collar. This collar is known in the trade as a Judas collar, and the donkey who wears it a Judas donkey, because she then moves on and finds another group of donkeys. The helicopter flies in, all these donkeys are killed, and the Judas donkey is again alone...until she meets up with the next herd. After this happens a few times, the Judas jenny stops looking for friends. She has worked it out: wherever she goes, the others die.

No one who has ever known a donkey could be unaffected by this practice, because donkeys are gentle creatures. Jill Bough found that the shooters were often sad about what they had to do, taking pride only in doing it cleanly. Most of us close our minds to the atrocities that we perpetrate on animals. That is the only way we can carry on as the dominant species and still sleep at night.

Our blindness towards animals reminds me of the way in which colonisers viewed those they colonised. Even the terminology is the same: brutes, dumb animals, uncivilised – meaning that they don't live like us, don't have the finer feelings that we have, don't speak our language. We have armed ourselves with prejudices against animals the way we once armed ourselves against indigenous peoples and women. The thousands of academics who have found in animal studies a new home may well be at the forefront of the next revolution, a new deal between humans and animals.

Anne Coombs is the author of *A Dog's Life* (Coombs, 2008). Her other books include *Adland* (William Heinemann, 1990) and *Sex and Anarchy* (Viking, 1996), and her essays have been published in *Griffith REVIEW: Webs of Power, Family Politics, Hidden Queensland* and *Participation Society*.

'Singer is far from the world's only serious thinker about poverty, but with *The Life You Can Save* he becomes, instantly, its most readable and lapel-grabbing one.'

NEW YORK TIMES

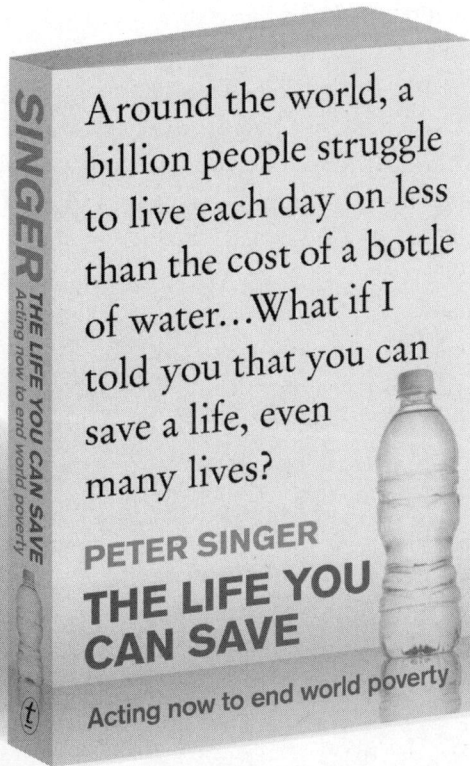

SINGER

THE LIFE YOU CAN SAVE

Acting now to end world poverty

Around the world, a billion people struggle to live each day on less than the cost of a bottle of water…What if I told you that you can save a life, even many lives?

PETER SINGER

THE LIFE YOU CAN SAVE

Acting now to end world poverty

'This is what philosophy is for.'

ADELAIDE REVIEW

AVAILABLE NOW IN ALL GOOD BOOKSHOPS

Text Publishing, Melbourne Australia

WWW.TEXTPUBLISHING.COM.AU

MEMOIR

In the apple orchard with Win and Petal

Postcard from a home producer

Melissa Sweet

ON the final evening of our week away, I took a plastic bag bulging with broad beans out to the veranda of the holiday house. There in the twilight I set to the fiddly task of separating the pale fleshy bodies from their fibrous green envelopes. As I dropped the beans into a metal bowl, my thoughts drifted to questions that would have made no sense at all to generations past. I can't imagine that my great-grandmother or her friends would have stopped to contemplate the merits of growing and making their own food. It was simply a matter of necessity.

As the bean skins began to pile by my feet, I wondered why so many people are returning to growing their own produce, whether in backyard, community, guerrilla or tree-change gardens. It's surely no coincidence that this is happening as supermarket shelves buckle under a dizzying range of foods, or at least products pretending to be food.

I make my living by writing about public health, and have been among those promoting home-grown and local produce (and encouraging people to avoid the processed pap on supermarket shelves) for the sake of planetary and personal wellbeing. I am inspired but also intimidated by the example of people such as the American writer Barbara Kingsolver, whose personal experiment in local eating and growing is described in *Animal, Vegetable, Miracle: Our Year of Seasonal Eating* (Allen & Unwin, 2008). She reminds anyone who might be tempted to romanticise the past of the slog involved in reverting to the ways of previous generations. I am not made of such stern stuff, and have no ambition to do the hard yards to become truly self-sufficient. I may make my own fetta, yoghurt and bread, but that's because I've discovered that it's not nearly as difficult as those selling 'convenience' foods suggest.

My motivations are nowhere near as high-minded as Kingsolver's, and have more to do with some primal instinct than with saving the planet. I dig and plant and grow because I must. And also because of my inner Scrooge. There are other reasons, too, but as I pulled apart the beans I realised that some of the way I choose to live is a throwback to the days when frugality was a necessary virtue.

PREVIOUS PAGE: *Untitled.* Photographer: Grant Harvey
from the *Sydney Morning Herald* Shoot the Chef! 2009 photographic competition.

PART OF MY preparations for going away was a harvest. As well as the beans, I picked silver beet, parsley and lettuce. I particularly enjoyed severing the broccoli stalks, after months of watching the plants grow bigger and leafier without any promise of flowering. Maybe they finally sprang into action because of my threats to consign them to the chooks. I doubt that it was because of anything I did; I am not a particularly meticulous or methodical grower. I put some effort into the earth – my excursions to town are judged successful if I find bags of horse or alpaca poo for sale by the road on the way home – but have a fairly relaxed approach to the plants themselves. Whether you call this letting nature take its course or survival of the fittest, it is amazing how much thrives on neglect.

As I packed my harvest, I wondered whether taking vegetables on holidays might be a sign that I really am going potty, a question that seems to be recurring with increasing frequency as my way of living moves further from the urban mainstream. In the end I decided that BYO produce was not nearly so silly as leaving the vegetables at home and then having to buy limp, inferior produce of unknown heritage. The broad beans were worth every ounce of effort; they were delicious tossed through pasta with the juice of a lemon from a tree at the holiday home. Yet I was not completely satisfied. It seemed such a waste to throw away that pile of skins. They could have been put to use if Scrooge's helpful friends, the chooks, had come on holidays too.

We arrived home to find Chris, a geologist, and Baina, an artist, bursting with news of their local explorations. We first met them a fortnight before our holiday, when they came for lunch and a mutual inspection after hooking up with us on a house-sitters website. Over home-grown spinach pie we discovered many connections, and by the time they returned for house- and animal-minding duties, Chris had much to tell us about the history of our area. As he talked, we began to understand another way of seeing the land around us. Where we saw the results of our labours over the past several years, he saw how rocks had been compressed, crumpled, cracked and cooked three to five hundred million years ago. Where we saw old farm buildings clustering in a clearing atop a bushy ridge, he saw an ancient river carving its way through time and space. Where we saw rich basalt soil nurturing our hundreds of plantings, he saw reminders of violent volcanic eruptions.

The visitors also gave us a much more recent picture of our property's past. They left a print propped on our old kitchen table that they had enlarged from a black and white photograph taken in the 1920s. The photographer had been standing on the hill below where our house now rests, and pointing the camera towards land that had recently been cleared of massive ironbarks, stringy-barks and other virgin bush. The settlers often took a week to fell and dig out a single tree, and the clearing went on for years before the apple trees lining the foreground of the photograph could be planted. Young and bushy, they stood in awkward contrast to the dense, unruly thicket in the photograph's background.

Chris and Baina found the picture when visiting our next-door neighbour, Win. Her parents began clearing the bush on our place in 1912, and subsequently planted 2,500 apple and pear trees. Win worked on the family farm after she finished correspondence school in grade six, until she married and moved next door in 1952. She remembers spending three weeks with two horses ploughing the paddock around the apples, and taking the harvest by horse and cart to the nearest village. The return journey took three hours; these days it is a thirty-minute drive.

Win is a fine advertisement for the fruits of hard labour, and is still collecting eggs, digging potatoes, harvesting fruit trees and chopping wood, despite having celebrated her eightieth birthday some time ago. She is one of the joys of our tree change. We are connected not only by place, but by a shared love of growing and animals.

LATE IN THE winter of 2006, Win came by to help us plant an apple sapling on the hill overlooking the paddock where her parents' orchard once stood. A terrible bushfire destroyed it in 1965, together with the other orchards that once made this area famous for its apples. Her father never recovered from the loss of a life's work, and died a few years later.

The tree we planted together, in honour of Win and her late husband, Harry, is called Scarlet Nonpareil and produces sweet red fruit. The heritage-apple producer from whom we sourced our fifty old-fashioned varieties describes this tree, whose origins can be traced back to around

1773 in Surrey, England, as yielding 'a very excellent dessert apple of first-rate quality'.

I could tell you that we planted heritage apples because we are part of a grassroots push to preserve the plants of the past, whose value exceeds those that drive commercial production. Or because sensible growers learn from history, and sentimental ones seek to repeat it. But really, we selected our trees for the poetry of their names.

Our orchard includes Autumn Pearmain, dating back to sixteenth-century England; Belle de Boskoop, whose lineage can be traced back to the Netherlands of the mid-1850s; and the Duchess of Oldenburg, whose popularity in Russia has lasted for centuries, thanks to a 'soft, creamy flesh'. Then there is the Ribston Pippin, 'one of the richest flowered apples'; the Beauty of Bath, known for its 'pretty and fragrant apple'; and Peasgood's Nonsuch, which is said to cook 'to a puree with a spicy flavour'. Not to forget the Boswell, which 'hangs well'. This is just a taste of the diversity that can be found in an old-fashioned orchard.

The trees had a tough start to life, thanks to the great wallaby massacre, but, incredibly, all recovered. Last December, we made our first tentative harvest from the juveniles. Into one bucket went 3.6 kilos of cooking fruit; into another went 1.3 kilos of eating apples.

Another few buckets were filled with fruit that had fallen or been eaten by birds or bugs. These buckets I stewed up for the benefit of Petal, our 350-kilo porcine companion, who prefers to take her fruit as dessert. Petal arrived in our household as a delicate week-old runt – thanks to Win – and captured our hearts immediately. She began her residency in a box before the fire in the lounge, then moved to a larger box in the laundry, and then to a small yard in the old apple-packing shed, before making her final move to an old set of cattle yards, where she now spends long hours renovating her mud spa. A few months ago she celebrated her second birthday by devouring an entire fruitcake in a few happy chomps.

When you have the pleasure of getting to know a pig well, you realise how poorly we treat the species, not only because of the inhumanities of intensive livestock production, but also because of the pejorative vernacular. Calling someone a pig is never a compliment, and yet I've rarely met a nicer creature.

Pet — by name and nature, as she has become — is full of humour and mischief, curiosity and affection. She can be naughty, unerringly making a beeline for the potato patch on our daily walk, knowing full well that it's out of bounds but remaining determined to get in a few quick munches before I catch up. She can be infuriatingly obstinate, even fierce, when it comes time to return to her pen. If you weren't well acquainted, the angry head-waving and tooth-baring might seem intimidating. But this subsides as soon as I put on a show to rival hers — waving my arms and growling threats — and our ritual confrontations invariably end with a friendly cheek or belly rub.

On meeting Petal, visitors invariably do two things. First, they state sagely that pigs are very smart. No argument there. And then they make tacky jokes about bacon.

I don't mind a bit of bacon, and have inadvertently ended up with a freezer full of free-range pork due to a mix-up with a local producer I found through the local Slow Food Association. I am sure he told me over the phone that the minimum quantity of pork available was ten kilos. But when I went to collect, a forty-kilo box was waiting. Instead of shelling out $150 — and that seemed such an extravagance — I was $600 out of pocket. Pork will dominate our menu for many months. I'm not sure that this is exactly what Carlo Petrini, the charismatic Italian founder of the Slow Food movement, meant when he told an audience during his recent visit to Australia that we should all be prepared to pay more for good food.

Petal, however, is safe. We would no more eat her than any of our other dear friends. No doubt, there are some who would take a dim view of this anthropomorphic approach. Such a waste, they might mutter. To which I reply that Petal earns her keep as part of the recycling loop: vegetable scraps into pig, into poo, into compost, into veggie patch. The Scrooge in me loves the pig, even though she is far from cheap to keep.

I HAVE SYMPATHY for the mutterers. Our lifestyle may be frugal by some standards, but by others it is incredibly self-indulgent. Few on the planet are able to live in such natural luxury while maintaining a semblance of a city job.

When I look out over our green hills, whose largest use is to keep the roos, wallabies, wombats and other wildlife well fed, I sometimes hear other mutterings. Bloody tree-changers, they grumble, making hobbies of what was once productive farmland. Once we brought a tall, gentle man from Nigeria here for a recuperative spell away from the trials of asylum-seeking. We took him for a walk in the bush along the back of our property, expecting he would be delighted by the friendly wildlife, which generally impresses visitors. He was impressed, but not in the way we had expected. He charged after the wombats, roos and wallabies, doing his utmost to knock them out with the large rocks. He thought us awfully wasteful to have all these animals around without making use of them. We were pleased his aim was bad.

One morning, I took a draft of this article for a short drive down a damp country road. I also packed a luxurious freshly picked head of broccoli. One of the rituals of visiting Win is that you never arrive or leave empty-handed. Win checked that the historical facts were all correct, and sent me home with a large slab of double sponge, baked in her wood-fired stove, and two packets of peanuts which had been bought for a Christmas long past and only recently rediscovered. She knew that Petal would appreciate them, even if they were a bit stale. My inner Scrooge, it need hardly be added, was most content. Frugal living fits very comfortably with abundance, after all.

Melissa Sweet is the author of *The Big Fat Conspiracy: How to Protect Your Family's Health* (ABC Books, 2007). She was awarded a Dart Centre Ochberg Fellowship for *Inside Madness* (Pan Macmillan, 2006), and is co-author of *Ten Questions You Must Ask Your Doctor* (Allen & Unwin, 2008) and *Smart Health Choices: Making Sense of Health Advice* (Hammersmith Press, 2008). She holds honorary appointments in the Sydney School of Public Health at the University of Sydney and the School of Medicine at Notre Dame University, Sydney campus. www.sweetcommunication.com.au

FICTION

THE SECRET LIFE OF VEAL

NICK EARLS

SHE struck me as attractive in a certain way, the way the head of a well-made axe can be attractive, hard angles matched to a purpose. She talked about modelling, though she never called herself a model. She said her name was Destiny, though when I finally saw her student card it said Carmel and had a picture of someone less harsh who she had once been or at least resembled. She told lies, but I knew she would, and I told plenty myself.

I met her at an art exhibition in a condemned house in New Farm, near the park. I had seen something about it on a flyer on a power pole and my run of luck had been predictably terrible, so I thought I had nothing to lose and everything to gain. I was recently single and had come to accept that I deserved every bit of it, with the exception of the irritating behaviour of my parents, who never stopped making comfort food, who kept Norah Jones on repeat and who told each other, more often than they needed to, that her music 'transcends generations'. My mother's phrase, and our first fight after I moved back in with them.

The art-student crowd — they're not fussy. That's what I was thinking. I might score some time with some vis-arts emo whose self-loathing was getting in the way of better judgement, and my mother could meet her over breakfast and that would somehow amount to a justifiable slap in the face for remaining my ex's friend on Facebook.

'Did you see what Maddie was saying on Twitter?' my mother had said the previous night, before she could suck it back in.

No. No, fuck you, no. I didn't see what Maddie was saying on Twitter. Didn't see or care, though of course I cared. Maddie could be so clever in a neatly streamlined 140-characters-or-fewer way that regularly left my mother marvelling. Marvelling at how smart Maddie was, marvelling that she'd stuck with me all those unrewarding months.

When it was over, my mother said it would be wrong to take sides. She gave me a house key and my old room back, and sent Maddie texts saying things like 'hope yr ok' and 'u gave it yr best'. I wanted her to take sides. I wanted her to take mine.

I had a point to prove, and only reckless coupling for shallow reasons would prove it. Even if I wasn't entirely sure what the point was.

The house where they held the exhibition was an old Queenslander that had risen from its stumps at the back, as if it knew its time was up and needed to break free before the movers came to take it off on a huge truck to one of those sad boneyards where unwanted timber houses go. There was a bedroom at the rear, suspended above the sinking stumps and inhabited in a very temporary way by a backpacker known only as 'the German'. He paid no rent but put in twenty dollars a week towards utilities, on the condition that he never use the landline. Which was fine, since he could siphon an unknown neighbour's wireless broadband with ease and have long insincere conversations with his girlfriend in Dresden on Skype.

Destiny had dropped her boyfriend that night, when he failed to pull his weight blowing up balloons for her 'installation', which was a room full of balloons. That had sent the relationship to the brink when the room had been filled only thigh-high, and then he pushed it over the edge by pulling out his mobile and ordering a meat-based pizza.

I tried not to salivate as Destiny described it to me. It had arrived on a Vespa — family size, meatballs, prosciutto, several kinds of sausage — and she recounted each ingredient as though it was a personal slight.

'I'm a very active vegan,' she said, with a hint of an American accent that I didn't quite believe. 'I have almost four hundred followers on Twitter. That's how Twitter can work for a vegan, if you have the heart for it.'

I told her I was one too, explaining that my belt was a very convincing leather-look synthetic polymer, and describing some fights I had never had with my parents when I supposedly renounced meat in a particularly sanctimonious manner at the age of twelve. Meat. I wanted to eat that pizza. I wanted to eat meat in an instant, powerful and caveman-like way. Prosciutto, meatballs, sausage: I was sure I could still smell them.

'If I could change only one thing,' she said, 'it would be veal. Veal is appalling. Veal is holding babies in a trap and feeding them milk and then eating their poor little muscles that have never become anything.'

She was like an actress playing an actress, and that was enough for me. An actress from the golden age of Hollywood, when women were admittedly bustier than she was, but wry and archly manipulative.

They had wiles in those days, which people don't so much now. She smoked too, like a lot if not all of the original Hollywood stars, though I had to admit she smoked less like Rita Hayworth – who had a famous upward exhale, displayed to good effect in *Gilda* – and more like an old man forcing sludge out of his lungs at dawn.

My mother would hate her, just enough. I could see the text to Maddie: 'omg u shd c what the cat dragged in'.

I first saw Destiny as she disdainfully appraised another artist's installation, which featured a genuinely sleeping friend in a cocoon macraméd out of shredded Woolies shopping bags.

'Some things,' she said with no particular feeling, 'don't mean shit.'

I liked that thought. I wondered if the artist's friend had been woken by her voice, but Destiny was already on her way out of the room and I found myself following her.

'Oh, did you think I was talking to you?' she said. I could make out a hint of an unkind smile. 'I was just talking.' It was almost dark in the hallway that ran down the middle of the house, some dim light sneaking in and making shadows in the grooves between the wall's vertical boards. 'I might talk to you,' she said. 'If you'll annoy me less than these people.'

I expected – knew, even – that I would annoy her just as much, given time, but for then I kept following. We went to the bathroom, where she walked up to the ice-filled bath, took the best bottle of someone else's wine that she could find and poured us each a plastic cupful.

'We did this with a grant, you know,' she said, looking at the fairy lights wrapped around the shower rose and flashing like a Christmas tree. 'From Arts Queensland. It'll all have to mean something when we acquit it, I suppose.' She drank a mouthful of the wine, and looked down into the bath, at the Vodka Cruisers and six-packs of beer, their cardboard packaging soft under the ice and coming apart. 'I loathe my boyfriend,' she said to the bath. 'My now *ex*-boyfriend. I loathe his hedonistic empty-headed conservatism.'

He was three ways bad, all in the one taut sentence, with more to follow. She told me about him there in the bathroom – his stupid remarks about not liking the taste of the red balloons, the offensive dead-animal pizza.

'You make it sound like road kill,' I said to her and she said, 'All his pizzas are about killing.'

And that led her on to veal and her grand plan for liberating veal, for busting into the feedlot and setting all the wobbly little calves free so that they could muscle up and not be eaten. 'You can be my lieutenant,' she said, and I imagined us going into battle together, and the sexually charged two-person victory party that would inevitably follow.

She was smoking by the time we got to veal, and further into someone else's wine. We were in the backyard, in a square of veranda light, with her artist friends around us in the dark, all talking about the shortcomings of the art of others, and people they shouldn't have slept with, and cafés they hated working at and the pretentiousness of the customers. Someone was filming much of it, and I wasn't certain whether they were being postmodern or just irritating.

My new great ambition, I realised, was to be the next person Destiny regretted sleeping with, to take my place in this life she obviously loathed, and to be one of its more loathsome parts. I was quite confident I could do that. I loathed myself exactly the right amount.

'We've got disaster written all over us,' I said to her. 'I'm in.'

She took my phone and put her number in it, spelling D-E-S-T-I-N-Y aloud as she did so.

'Saturday, early, your car,' she said. 'I'll text you the address.'

I HAD A week to plan, to veganise as much as I could. I googled veal; I bought a hideous belt that only a vegan could love; I attempted unsuccessfully to track Destiny down on Twitter, in the hope of picking up some of the nuances of her anti-carnivore rage. I practised eating vegetables. Didn't like it much.

I burned CDs of vegan music for the car, or at least music by vegans. At first I thought that'd mean a dangerous amount of Coldplay, on the strength of Gwyneth Paltrow's assiduous compliance with a macrobiotic diet. Which was quite like vegan, surely, or close enough. But then Wikipedia saved me with a list of vegan musos, and I had The Church (Steve Kilbey), The Smiths (Morrissey and Marr), Michael Franti, Kisschasy, Jonathan Richman for retro cool, Moby for highway driving, some Fiona Apple for when I wanted to look clever, Radiohead for when I wanted to look that kind of clever and k.d. lang for closing music, though I wasn't sure if it was just lesbian closing music. We'd park somewhere at sunset, calves gambolling across the paddocks, and I'd crank up 'Constant Craving': welcome to the victory party.

By Saturday, I had two CDs chock-full of vegans, without a Bryan Adams or Shania Twain track in sight. Not just vegan music, but discerning vegan music. The music of a vegan you would probably want to have sex with.

I borrowed – or in fact, took – my father's new car since, unlike mine, its upholstery wasn't tainted by years of delicious flame-grilled burger smell from a life of miserable broken-hearted car meals. I left him my keys and a note telling him I was sure my car would do, for a Saturday.

Destiny was standing in front of the rather grand house at Indooroopilly when I arrived. She looked not so much like a resident and more like someone who had just stepped off a bus.

'Bye,' she shouted back to no one at all as she got in the car.

The stereo was playing the Indigo Girls' 'Closer to Fine'. Destiny checked her nails, first on her left hand, then on the right. The Indigo Girls played with an energy that belied the lack of quality haem iron in their diet.

Destiny unzipped the bag she'd dropped at her feet and said, 'You're right to use bolt cutters, yeah?'

'Yes,' I said, meaning no and that's altogether too hardcore and did you notice that's the Indigo Girls and at least one of them is vegan and you probably want to have sex with me now?

I had seen bolt cutters in movies, wielded by bad tough men. I didn't know there would be bolt cutters.

'Have you ever had coffee at Extract, on Adelaide Street?' she said after a while, making it sound unlike a question. 'I recommended it to Lily Allen when she was in town.'

That actress voice was back, as if she were playing a jaded southern belle. She left a pause there for me, which I thought I was to fill with some observation about her being Lily Allen's friend.

'I don't play much Lily Allen,' I said. 'I don't mind her forthrightness, but I think she eats meat.'

I turned the stereo up a notch, then down, to draw attention to the music of Conor Oberst. Please note the cerebral alt-folk vibe, I was willing her, and the complete absence of animal products.

'It's not as if we're close,' she said. 'It was on Twitter. She did ask, though. She didn't reply. Not that I noticed.' I asked her how many people followed Lily Allen on Twitter and she said, 'About a million and a half.'

We drove in silence for a while. I worked on vegan thoughts, but none would come.

'My uncle has a lazy eye too,' she said. She was looking at me, watching me, and turned half sideways in her seat. 'A slightly buggy, lazy kind of eye. He's got one of them too, so I shouldn't worry. Have you ever watched any of the films of Marty Feldman?'

'I don't think so,' I said, suddenly aware of my eyes, pushing out of their sockets like a pair of fists. Which they weren't. Just plain weren't. No fists, no pushing. I had a tendency, perhaps, to look a little more surprised than I was, but symmetrically, damnit. And only slightly. My eyes felt like big balls of dense jelly in my head, lids holding them in like seatbelts. She had found me a new way to be ugly, and I already had more than enough of those.

'Doesn't matter,' she said, still looking at me closely, and not in a friendly way. 'My father made us watch them. It's a kind of abuse.' She laughed. 'It's not abuse. There was no abuse in my family.'

We headed west out of the city, and then north, past the half-empty dams and through towns selling pies and dry hopeless land.

'I can draw a perfect circle in real life,' she said, and I wondered if there were other places to draw circles. 'By hand.' It was added as a kind of correction.

She made the shape of a circle in the air in front of her. It may have been perfect.

'My mother has always wanted me to be someone else,' I told her. 'And I tried, but I could never work out who else to be.'

'That's completely pathetic,' she said. 'Not just pathetic, but completely. I'd like to write it down. Do you have a pen?'

I didn't, so she reached for her phone and spelt it out aloud, letter by letter, as she typed it, vegan music drifting meaninglessly by, not in the same country as sex.

She navigated us through the Brisbane Valley using a map she'd printed out, first along the highway and then down country roads, past dry bush with empty bathtubs and burnt-out cars, past lush horse studs with pokerwork signage. We left the bitumen and moved onto a graded surface, and then something less kempt and more like a track.

'I'm sure we're closing in,' she said, as a tyre blew and my father's car, which had never before left the city, fishtailed in the dirt, throwing up dust and rocks and splinters of tree branches.

We came to a stop at the edge of the cleared surface, in the ruts that some large wheels had made when the ground had once been wet. The dust cloud we had thrown up blew by us.

'Your music is sometimes irritating,' she said, and turned it off.

I meant to shut the door when I got out, but it slipped from my hand and slammed instead. I went to open it, to explain, and then didn't. She was staring straight ahead, waiting for the car problem to be fixed.

The tyre had shredded and left chunks of rubber back along the road. The day was hot now, and close to silent in this lost bit of country. I was wrecking my father's car, veal would stay veal and I was not even going to have sex.

I found the spare and lifted it out of the boot. I made sense of the jack and pristine toolkit, as much as I could. I fitted the heavy iron wrench to a wheel nut and pulled back on the long handle. Nothing. No movement at all. I pulled harder, jerked at it. Something clicked in my neck, but the nut didn't shift.

I repositioned the wrench and stepped on it. I put my hands on top of the car and lifted all my weight onto the wrench and stood there, several centimetres above the ground.

I stepped down, and stamped on it. Then stamped on it hard, and the heel cracked off my boot and a pain shot up my ankle and I fell over. The wrench slid off the nut and fell onto the dirt.

Was there any prospect of sympathy sex? Was it all I had left?

I stood up again, a limp now built into my shoe asymmetry. I opened the car door just as Destiny was lighting a cigarette. I wanted to tell her not to. I wanted to tell her the day had gone to shit in more ways than I could count. I wanted her to like me, or something.

'What?' she said, indignant just for the hell of it. 'You're not going to tell me there's no smoking in the car now?' She drew on the cigarette, glared and opened her window several centimetres, as if it was a compromise. 'Some people are actually allergic to new-car smell,' she said, because I was in the wrong again and the day was shit in an extra way that I hadn't bargained on. 'It makes them sick. They can't buy new cars.'

I had never owned new-car smell. I had borrowed it, and then covered the car with red dust, shredded a tyre and scraped the passenger side against roadside bushes.

'I read it on the internet,' she said, less robustly.

I had stared her down by accident. With the pain in my foot and the private tallying of the ways the day had gone to shit and my simple inability to muster a reply, I was accidentally coming across in a masculine, no-nonsense kind of way.

'Well, at least it comes from an incredibly reliable source,' I said to her. 'The same site where you found out Jeff Goldblum's dead, right?'

That made her laugh, which was an improvement, though not in itself a reason to turn the sound system back on and cue k.d. lang.

A truck appeared around the bend ahead of us. An empty cattle truck, coming our way. It clattered and rattled, all of its parts that were normally separated by cattle thrown against each other with every bump in the road. Dust blew out beside and behind it in a pink cloud. I stood up and shut the car door, tried to make some kind of signal that non-emergency assistance was needed.

With a rush of air from the brakes and more clattering, the truck stopped some distance in front of us. The door opened and the driver climbed down. He wore an old checked shirt with the sleeves torn off, shorts that had done a lot of miles and boots that looked like the kind with steel caps. He was fiercely stocky and fortyish, or younger and weather-beaten, and he walked as if he habitually sat on a haemorrhoid cushion.

'You'd be a bit off the beaten track then, wouldn't you?' he said when he got closer.

'We heard there was a winery out this way.' Even as I was saying it, I knew I could have done with a better lie.

'Winery?' he said, and laughed, as if I'd said milliner, or American Nails franchise. He was looking at me as though I had no penis at all. 'Winery. You and your lady up for a nice chardie from these parts, are you?' He smelt of cows, and therefore cow shit. 'Tyre blowing chunks, then?' he said, nodding in the direction of the problem.

'Yeah.' That kept it laconic, and as manly as possible.

'Righto.'

He walked past me, picked up the wrench as if it was no heavier than an Allen key, and hoicked off the first of the wheel nuts with a flick of his wrist.

'What's this?' he said, picking up the broken heel of my boot. 'Take a bite out of your shoe, did it?' He flicked off another nut. 'You might want to slip the jack under and think about cranking it up a bit.'

I was as manly as an orchid to him. He had the last two nuts off before I'd found the hole for the jack handle, so he took that job over as well and had the tyre changed in minutes. Destiny sat in the car and continued to smoke.

'Good luck with the wineries,' he said, handing me the jack.

'Yeah, no worries,' I said back to him, though it sounded stupid. 'Thanks for that.'

He was already on the way to his truck, and he waved without turning. He climbed aboard, gunned the engine into life and gave a couple of painfully loud blasts of the horn as he passed us, sitting far above us in his air-conditioned cab, lost behind the dark window tint.

I opened the car door.

'Maybe you could stub that out for me on the road,' Destiny said, handing me her cigarette butt. 'If you think you can do it without getting hurt.'

AGAINST MY BETTER judgement, we drove on. Destiny still had her map, and said we were close, and that the cattle truck proved it. We took a left after a dry creek bed, then a right, and all of a sudden the rank smell of a thousand cattle held at close quarters came in through both open windows.

We passed the entrance to the feedlot without slowing down. There were men in the yard, all like the truckie and able to snap me like a pencil, and No Trespassing signs in the sternest possible font.

'Shit,' she said.

We took the next left down a rutted lane and pulled over under a tree. I wondered exactly how heroic I needed to be for a sexual outcome. Those men would hurt me badly even if I squealed and said it was all her idea.

'Right,' she said, and handed me the bolt cutters.

I could see the feedlot goons taking them from me, clipping off a digit or two and laughing. Tossing me into a mincer, feeding me to pigs. I had watched too many of the wrong movies.

'We need evidence,' she said, and took a Handycam from her bag. 'This has to go on YouTube. We'll pixelate your head, or something, if it ends up in the shot. You go first.'

I stepped out into the rank cow smell and pushed through the bushes, towards the mooing. After a few metres I could see clear ground ahead, and cattle, and then I made out the wire strands of the fence.

'Cut it,' she said. 'Cut it.' And I did, and then she said, 'It might have been electric. Was it electric?'

The cattle watched idly as, successfully unelectrocuted, I cut the middle strand of wire, and then the bottom one. They were crowded into the paddock, with the feedlot buildings beyond them. Any grass had long ago been chewed away, and they were standing in dust and generations of shit.

'Okay,' she said. 'Now we set them free.'

She videoed me approaching the cows, bolt cutters in hand, limping on my broken shoe and waving my arms like a silent-movie fool, a Keystone Kop, miming some kind of urgency around the theme of departure. The cattle backed away, closer to the buildings.

'You stupid fucks,' Destiny said, one hand waving at them, the other filming. 'That's certain death.'

We chased around for twenty minutes and the cattle kept backing away, did whatever they could to steer clear of us. If we herded one up and chased it to the hole, it veered off at the last and jogged back to the others.

'Fuckers,' Destiny shouted, then grabbed at a tail and fell in shit. Actually, several shits. One mid-thigh, one on her front, one on her left ear and strands of hair. 'Fuck you!' she screamed at the perplexed cow. 'You totally fucking deserve to die.'

It chewed on something, perhaps a memory of past cud. It blinked.

She slapped its face and said, 'Don't look at me that way.'

We heard a motorbike engine up near the buildings, and ran.

'You have to give me your shirt,' she said once we were back at the car. 'Don't argue. Just give it.'

I stayed on the other side of the car, discreetly pumping my bare muscles in the hope that they could look half-respectable, while she swore and moaned and dry retched and used her own ruined shirt to wipe shit from her hair and jeans before throwing it under a bush.

'You can turn around now,' she said. She looked good in my shirt, albeit pale and queasy and angry, and with a chunk of her hair stiff with drying shit and standing out from her head like a blade.

'Well, there goes the last of the new-car smell,' I said as we got in, and she said, 'Fuck you.'

WE DROVE OFF with the windows down but the smell of the byre was inescapable and not improved even by the best vegan music in the world.

'I thought you didn't wear animal products,' I said, and she said, 'Jesus Christ. I'm only just not screaming.'

There was a McDonald's on the edge of the first town we came to and she said, 'Pull in. There'll be a tap. They have to have an outside tap to maintain their shrubs.'

I did as I was told, and parked in a corner away from the drive-through. She found the tap and went down on her knees to wash her hair and her jeans, and shriek at the neat tan bark, at the horror of it all. Shit ran from her hands. Shriek, scream.

She came back to the car drenched, my baggy shirt stuck to her lean body. I filmed her with the Handycam, and she gave me the finger.

'Fuck it,' she said when she was no more than a metre away. 'I'd kill for a burger. I want to eat animals and smoke more to keep my weight right.'

She found new ways to be vile at every turn. My parents would hate her, if they ever got the chance.

'You know that was my art?' She was pointing at the Handycam, and I was watching her through it, still filming. 'You know that, don't you?' she said, and it was plain that I didn't. 'That the balloon story was all bullshit and my art was about pretending to be a vegan and seducing some poor fool with that horrible story about veal. Which still icks me, by the way. We shot the whole thing. My friend had that camera. It's already on YouTube. That's the art. You being a tool. A totally dick-led tool. You realise you're the proof that the worst things said about men are true? That you'll agree to anything if there might be sex at the end of it?'

I told her I had a voucher that would get us two steaks for the price of one at the Indooroopilly Hotel and she said, 'Take me there. Stat.' She got in the car, wound down the window, lit another cigarette and said, 'I think I'm going to have sex with you after all.'

And I said, 'Yes, at my house. And then you'll meet my mother. You can smoke and wear my clothes and still smell faintly of shit, right in her pristine kitchen. I'll make her cook Wiener schnitzel, and that's when you can tell her all about veal.'

Nick Earls lives in Brisbane and is the author of thirteen books of fiction, including *The True Story of Butterfish* (Vintage, 2009) and *Bachelor Kisses* (Viking, 1998). His memoir 'Songs of childhood' was published in *Griffith REVIEW: The Lure of Fundamentalism*.

ESSAY

Beyond the recipe

Food writers as activists

Donna Lee Brien

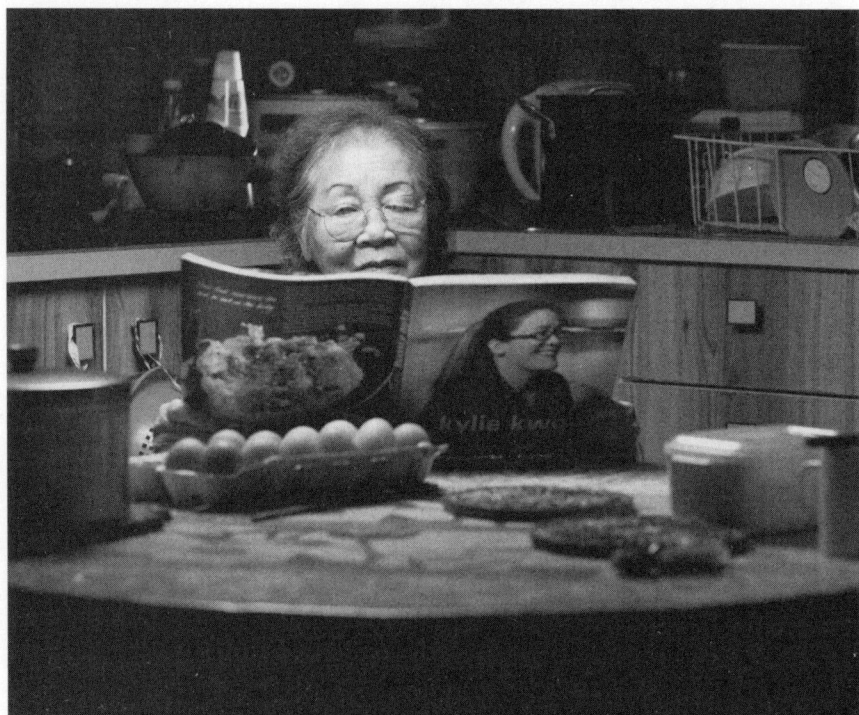

'**EATING** is not merely a biological activity, but a vibrantly cultural activity,' the food anthropologist Sidney Mintz reminds us. Our eating and shopping choices are now frequently imbued with complex moral choices. Should we, for example, be buying locally produced foods, with all the benefits for our own region and our health that this may provide? Or should we be choosing Fairtrade goods, with all the benefits for producers in the developing world, but with a much larger carbon footprint from transportation?

The associated questions – of whether Fairtrade is actually fair, and what exactly 'local' means – can lead to similar impasses. The idea of local can vary from twenty to two hundred kilometres, and some people define it as a region or even a country. Such discussions slide into food politics: the ethics of food production and consumption and, in particular, the sustainability and viability of food systems. This politics is being shaped by figures outside government: chefs and farmers, environmental activists and concerned consumers, as well as all kinds of media producers, publishers and writers – especially food writers.

Food writing includes recipe and cookbook writing, food journalism and restaurant criticism for magazines and newspapers, books of food history and anthropology, and food-focused biographies, memoirs and essays. Travel writing often crosses over into discussing food, as does scholarly writing on nutrition and health, science and environmental issues.

Fiction, too, is in on the action: novels featuring recipes, menus and other cooking-related information are perennial favourites. There are restaurant-based romances, but by far the most significant – in sales numbers – is the subgenre of mystery known as the culinary cozy. Starring amateur detectives who are chefs or caterers, restaurant critics or gourmands, police officers or private detectives interested in food, these books not only contain richly written descriptions of cooking and eating but, when most successful, plots that revolve around food.

PREVIOUS PAGE: *Mrs Jang's eggs (detail)*. Photographer: Nicholas Williams from the *Sydney Morning Herald* Shoot the Chef! 2009 photographic competition.

There are online manifestations of all these kinds of writing about food, with foodie blogs attracting ever more attention. Last year more than two million people visited Australian food and cooking websites. Online and in print, hybrids are also common, such as cookbooks built around collections of poems or local histories, and novels, memoirs and blogs on non-culinary topics that include a high proportion of recipes and other food-related material. These permutations create interesting dilemmas for bookshop staff. Lizzie Collingham's surprise 2006 bestseller *Curry: A Tale of Cooks and Conquerors* (Vintage) can be found, for instance, in the cookbook, food essays and criticism, history, travel, biography, and home and gardening sections of bookshops.

Although the number of books sold in Australia has been relatively stable since 1995, with around 130 million sold in most years, books on food have been selling in ever larger numbers over the past decade. Nielsen BookScan, which logs most Australian retail sales, reports that in 2006, two million food and drink books, worth almost sixty million dollars, were sold. Food and drink titles make up a similarly considerable proportion of the Australian magazine industry's annual sales of more than one billion dollars. Of the six thousand magazine titles available, many are from overseas but, in terms of circulation, most of those in the top hundred are Australian. In 2007, there were four food magazines in the top-selling twenty, and eight in the top fifty; Australians bought six times more food magazines per head of population than the British. The flow-on effects to our culture are significant: revenue from the sales of popular cookbooks helps subsidise the production of texts with smaller readerships, such as literary fiction, poetry and history.

MARK KURLANSKY, THE author of *Cod: A Biography of the Fish that Changed the World* (Penguin, 1997), *Salt: A World History* (Random House, 2002) and *The Big Oyster: History on the Half Shell* (Random House, 2006), believes that writing about food is 'about agriculture, about ecology, about man's relationship with nature, about the climate, about nation-building, cultural struggles, friends and enemies, alliances, wars, religion. It is about memory and tradition and, at times, even about sex.' Yet the global food industry works with the entertainment industry to entice consumers to embrace a 'live to eat' attitude – where to eat is primarily an act of purchasing. In such a world, food writing can become what Warren Belasco has called the 'handmaiden of consumer culture'. Paul Hertneky has characterised readers of 'food porn' as 'lavishly endowed foodies [who]…gorge on images and words, rapturous words, stern words, clever words, words in the mouths of stars, experts, chefs and doctors, words off the fingertips of those like me, who obsess about food, unleash our imaginations on food, craving and coveting it, loving it and fondling it, very much fearing it, and essentially having it replace sex in our middle age.'

Television chefs in particular make each cookbook an extended advertorial for themselves and their range of branded products. Yet some of these celebrity chefs are also generating debate about important issues. Love him or loathe him, Jamie Oliver has influenced both public behaviour and government policy. His exposure of the abysmal lunches provided by schools pressured the British Government into spending an extra £280 million on healthy food for students. His exposé of the conditions in which battery hens are raised resulted in sustained increased sales of free-range poultry and decreased sales of factory-farmed chickens across the UK, as well as complaints that the increase in free-range chicken sales would be greater if producers could keep up with demand. In direct response, some British councils have banned battery-hen eggs from kitchens in schools, nursing homes, town halls and canteens.

A NUMBER OF prominent Australian food writers and TV presenters have been similarly concerned with matters beyond cooking advice. The issues that have particularly concerned them include the work–life balance, gender equality, sustainable and ethical agriculture, biodiversity and genetic modification, food miles and fair trade, food safety and security, and obesity, diabetes and other health issues. In these areas, they are not only media commentators on contemporary concerns but, at times, forward-thinking activists and campaigners for change.

Margaret Fulton's cookbook and magazine writing has, since the 1950s, both affirmed the importance of creativity to cooking and encouraged the use of convenient, time-saving products. In this, she has recognised women's changing roles and their increasing desire and need to work outside the home. Her writing has gone a long way towards making women's domestic work visible, and she has encouraged other women to take up meaningful work (even if it meant less time in the kitchen), both by positing that other career paths were possible and by her involvement in mentoring schemes.

Fulton is a fine example of the food writer as activist. In 2003, the giant multinational Kellogg's sought to patent in Australia the recipe for the chocolate-flavoured no-bake confection known as chocolate crackles, and have the trademark registered. In 1953, a trademark had been granted to Kellogg's for 'Chocolate Crackles', for what was described as a 'breakfast cereal term'. In the intervening fifty years, though, the chocolate crackle became a standard item at children's parties, fairs and cake stalls across Australia. Indeed, it is often included in folklore studies and surveys of iconic Australian foods, alongside the meat pie, Vegemite, pavlova, lamingtons, Arnott's biscuits and Aeroplane Jelly.

Legal advice suggested that this new 2003 trademark would mean that anyone selling chocolate crackles, or reproducing the recipe, would have to do so under the Kellogg's brand. The company would be entitled to payment, or a percentage of the profits from any sale. Newspaper articles noted this would mean that any cake stall with a sign reading 'chocolate crackles' next to a plate would be in breach of the new trademark; but the real resistance to Kellogg's claim was animated, and organised, by Australian food writers – led by Margaret Fulton, who wrote on the subject in newspapers and magazines, and participated in radio talkback programs. She drew attention away from the question of who invented the recipe, and focused on a statement by the government patent office, IP Australia, that any decision would need to consider the 'normal understanding of the words in the community', and whether anyone else would need to use the recipe to carry on their business.

At the same time, Fulton became involved in the debate about genetically modified foods, and she brought to her many readers' attention that in the US, shareholders were reacting negatively against Kellogg's plans to use genetically engineered ingredients in their products. Fulton became the face of Greenpeace Australia's part of the international boycott of Kellogg's products and, in the same year, 2003, she launched the second edition of a guide to non-genetically-modified ingredients, the *True Food Guide*, for the organisation.

I would not want to make inflated claims for Fulton's involvement in the legal situation, but a scan of newspapers, and women's and food magazines, from this time shows a significant increase in the number of articles about, or including information about, GM foods. These discussions included health concerns regarding allergic and toxic reactions, and possible increased resistance to antibiotics and cancer risks. Environmental problems were covered, including genetic pollution, the creation of so-called superweeds, and the potential increase in herbicide use and pesticide residue, as well as concerns over long-term effects on soil fertility, biodiversity and food security. Also discussed were the socioeconomic ramifications of food monopolies controlled by a few multinational companies. The debate had moved a long way from deciding who owned the recipe for chocolate crackles.

OTHER WELL-KNOWN AUSTRALIAN food writers of the past few decades – in particular, Beverly Sutherland-Smith, Peter Russell-Clarke, Stephanie Alexander and Maggie Beer – have promoted local, seasonal and fresh foods, and campaigned for improvements to the welfare of farmed animals. They have questioned the safety and nutritional quality of contemporary foods, especially highly processed products and those grown with chemical pesticides and fertilisers. They give voice to concerns about the social justice of sustaining rural industries and rural communities, in so doing discussing more sustainable and ethical ways of farming. And they are promoting resistance to supermarket-led food culture. Their guidance on resisting the big chains' marketing is often published in mainstream magazines, meaning it sits beside the very advertising it critiques.

One the most overt examples of Australian food writers' activism is the Adelaide Declaration. On 26 October 2005, a group of 'powerful' food professionals – including broadcasters, educators and producers – met at the inaugural Adelaide Food Summit to challenge state governments and other bodies to address the future of Australia's food and the health of its people.

The declaration called for access to 'good, safe and wholesome food' for all; government support for 'sustainable small-scale agriculture on the fringes of large population centres', as well as 'agricultural traditions – like organic farming – that strengthen biological diversity'; and schools to provide information on basic cooking, gastronomic and health issues. New food technologies, the signatories said, should be adopted with the utmost caution.

Johan Pottier, at the conclusion of his recent study of the social dynamics of food security, wrote: 'studying food issues, whether within households or in the offices of policy-makers, must not be "just academic". The aim of such research must be to understand and to transform [behaviour].' Australian food writers, by making us think about what and why we eat, and how it affects our world, are contributing to such discussion and understanding and, hopefully, such transformations of habits.

Associate Professor Donna Lee Brien heads the School of Creative and Performing Arts at CQ University, Australia, and is the immediate past president of the Australian Association of Writing Programs. The research for this essay was completed as a Research School of Humanities visiting fellow at the ANU.

Born in Vietnam, made in Australia

Getting the fish sauce recipe right

Pauline Nguyen

MY parents are known as members of the 'first generation' of Vietnamese refugees, who came to Australia after the Vietnam War. I, however, am known as part of the '1.5 generation'. Born in Vietnam, made in Australia. We are the children of defeated warriors who have tried to come to terms with the present life, and the act of negotiating the past with all its rules and traditions, in the hope that the two very different cultures could blend into one well-adjusted whole. This always seemed better in theory than in practice.

When Saigon fell to communist rule, in 1975, my father realised that he had no choice but to escape Vietnam. And the only way that he could do this was to build a boat and smuggle his family out to sea. I was three at the time and my brother Lewis was two. My grandmother begged my father not to leave. She couldn't understand how a parent could risk perishing at sea. But my father is a very determined man.

He stands at just five foot one, a little shorter than me, but what he lacks in height he makes up for in fearlessness and determination – and he had already made up his mind. He would rather die trying than risk imprisonment. Or a fate far worse, the re-education camps. 'It's not enough that they want to take our freedom,' he would tell me. 'They want to take our thoughts as well.' My father was determined that if we died, we would all die together.

So in October 1977, armed with only a rudimentary map and a compass, my father steered our tiny vessel out into the South China Sea. We spent days drifting and waiting and praying. We prayed that a foreign ship might come and save us. We prayed that we might find friendly shores. We prayed that the pirates wouldn't attack us. We prayed that our supplies would not run out.

Our prayers were not always answered. Ship after ship ignored our SOS, and at gunpoint a group of Malaysian soldiers pushed us off supposedly friendly shores before we landed in Thailand, where we spent a very difficult year in a refugee camp. Australia finally accepted us and put us up at the Westbridge Migrant Hostel, in the Sydney suburb of Villawood. My father quickly found a job working on the production line at the Sunbeam Factory in Campsie – on the graveyard shift from 2 pm to 2 am, the job nobody wanted.

The train ride home was the worst, he would later tell me. Every night was dangerous. The locals threatened to beat him and the worst bigots threatened to kill him. 'Go home to your own country, you bastard,' they would yell.

My father cried every day going home on that train. We all cried a lot in those days. We came into a new country with nothing: no job, no house, no money. We didn't know the laws, the language or the systems. My father had nightmares – the same dream, over and over. He's back in Vietnam, preparing for our escape. He's back in the water, drifting day after day with nowhere to go. And then he wakes up.

MOST VIETNAMESE WHO came to Australia during this time settled in either Melbourne or Sydney. My father chose Cabramatta, in Sydney, for its strong sense of community. He liked the idea that a number of his friends had already set up a life for themselves in such a short time. He understood that the secret to their success was hard work and unconditional dedication, often fuelled by underlying desperation. It didn't surprise him that many of them had become astute business people, showing great aptitude as shopkeepers.

In the mid-1980s, my father fed his sudden urge to open a video library. 'Asians love action movies,' he would say. The blockbuster releases at the time were *Full Metal Jacket*, *Platoon*, *Born on the Fourth of July* and *Good Morning, Vietnam*. My father found a prime location in the centre of bustling John Street, the spine of Cabramatta's thriving commercial centre.

'But why a video library, Dad?'

'Same as the driving school,' he said. 'No one else in Cabramatta's doing it.'

My father was considered a pioneer in the business community. Before opening a restaurant, he had dared to approach the local council about outdoor seating and shop renovation – a considerable feat for a new Vietnamese migrant. He became the first to offer alfresco dining, which proved a huge success.

Our restaurant offered a place for lonely migrants to meet and chat in their new and native tongues over a shared meal or coffee and ice-cream as they sat and watched the world go by. It made my father happy that his contribution and participation had helped so many rebuild their lives with confidence and hope.

Everybody works hard in Cabramatta. Seven days a week, the commercial centre is at full steam. Rest time comes once a year, in February, to celebrate the Lunar New Year. For most of their lives, my parents have worked seven days a week from 6.30 am until 10.30 pm. My father has closed the restaurant only once, when my grandmother passed away.

MY FATHER HAD constant flashbacks to the war. Part of his job as a lieutenant in artillery was to go back to the scene and count the dead bodies after a kill. One shell killed so many. The scars from his own bullet wounds resemble a question mark down the length of his spine.

Determined to succeed, my father took on a second job and then a third. At home he was always angry. He had an anger that none of us could explain. He would throw and smash things, and yell. Sometimes, he would just stand there and scream.

It wasn't long until he started to offload his anger on my mother, then on us, his children. My father was determined to raise four high-achievers. He wanted to make sure that the sacrifices he and my mother made were honoured.

If someone were to ask me what I remember most about my childhood, I would tell them it is overwhelming fear. Fear followed me everywhere, every day. My father kept three instruments of torture: a stiff and shiny billiard stick, a flexible cane whip and his most effective weapon, fear.

Twice a year we would bring home our school reports with dread. For every B, he caned us once. For every C, he caned us twice. We had to lie flat on our stomach and not budge until he finished, blow after blow hacking at the flesh on our buttocks and thighs. When he was done, he threw us a dollar for every A.

He used to say, 'I created you and I have the power to destroy you.'

At seventeen, I ran away and spent many years hiding from my father. I would look over my shoulder everywhere, paranoid that familiar faces might follow me.

THERE COMES A time when you need to conquer fear. For the sake of my mother and my brothers, and for all the shame I brought my family while I was away, I reluctantly reconciled with my father. I would go home to visit out of duty. I hated those visits. I hated the sense of claustrophobia and suffocation I felt in his presence. Our meetings were stifled, false and tense.

What I hated the most was the realisation that I had grown up to be like him. I too was angry all the time. Angry at my loved ones, my friends, my work colleagues; angry with the world. Angry at myself. Angry people are very skilled at noticing all that is wrong.

Later, when my partner and I decided to have a child, I was determined that this cycle would end. I was determined to not be the same person I had always been, because I was frightened, frightened of history repeating. Frightened of treating my own child the way I had been treated.

Towards the end of my pregnancy, I landed a book deal to write a memoir about my family. As I wrote, my fears returned. I worried: How could I possibly survive my father's reaction to the story?

There are ten chapters in my book. It's not meant to be a scathing account of my father's behaviour, but a beautiful story about personal freedom, family and hope. But in order to talk about the good things I had to talk about the bad things. I planned to finish the book and give it to him, so that he could see the full arc of the story. As I wrote, a cloud of dread hung over me.

By the time I finished the seventh chapter, my father demanded to read it. I freaked out. The seventh chapter was the most confronting, the most scathing about him – the most difficult chapter to write. I thought, *He can't read it now.* But you don't say no to my father: I had no choice but to hand over my unfinished manuscript. The story of his life, written by his prodigal daughter.

I didn't hear from him for two months. I needed to finish my book and move on, so on Father's Day I decided to go home and face the music. With my beautiful baby daughter, Mia, I drove home to Bonnyrigg to confront my parents. I was so nervous and scared I could hardly breathe.

I'M NOT SCARED that he's going to hit me; we've passed that stage, I'm scared because my writing exposes him and our family stories and secrets to the world. I'm scared because he might give me some ridiculous ultimatum and say, 'I forbid you to publish this book.'

I'm scared because I'm about to do something that's never been done before. I'm about to take responsibility to end this family's pattern. I'm about to confront my father to make things better.

So Mia and I wait at the front door. I've brought a case of my father's favourite red wine as a peace offering. When the doors open they take Mia, kiss her, cuddle her; they're so happy to see her. I see that they've made a feast for me. When we sit down to eat, I ask, 'Dad, what do you think about my story?'

'It's good, it's good, but there's just one thing wrong.'

'What's that, Dad?"

'The fish-sauce recipe's wrong.'

'What do you mean, the fish-sauce recipe's wrong?'

This can't be happening.

Later, I ask him again: 'Dad, what do you think about my book?' I get the same answer about the fish-sauce recipe. I'm frustrated that we're never going to define our relationship. We're never going to connect; I'm never going to finish my book; I won't be able to move on. I get Mia ready and gather our things.

I'm just about to leave when I ask him one last time. 'Dad, what do you really think about my book?'

And in a voice sad and serious he says, 'Do you know why Buddha sits on a lotus flower?'

'No, Dad. Why does Buddha sit on a lotus flower?'

'There is nothing as beautiful as a lotus flower. Out of watery chaos it grows. Emerging from the depths of a muddy swamp, and yet remaining so pure and unpolluted by it. So pure you can eat it, all of it, the leaves, the roots, the seeds, the petals. But the lotus flower has another characteristic. Its stem you can easily bend, but you cannot easily break. It has tenacious fibres that hold the plant together.

'My children are lotus flowers. You have grown out of the aftermath of war. You have grown up in Cabramatta during its murkiest time. And you have grown out of me. I am mud, I am dirt, I am shit. I am very lucky to have you all.'

With those words he gave me everything I had been waiting for. He never apologised, but he acknowledged the harm he had inflicted. Now, when I think about my father, I think about forgiveness; I think about redemption, and about hope, and about unfailing courage in the face of adversity.

IN OCTOBER 2007, Murdoch Books published *Secrets of the Red Lantern*. It has been translated into two languages, won numerous awards and touched the lives of many people. This made my partner, my brother and me realise that as restaurateurs, business people and human beings, we have a social responsibility to make a difference.

At Red Lantern we have embarked on a journey to promote ethical eating. We use the freshest in local sustainable and organic produce, and aim to leave as light an environmental footprint as possible, while staying true to our Vietnamese origins. We have transformed Red Lantern's backyard into a 'garden of tranquillity' where we grow our own herbs and vegetables. We recycle everything, even our food scraps. Our aim is to reduce our garbage waste by at least two wheelie bins a week.

WHEN I WAS asked to write a second book, my initial answer was no. I had to face many personal demons to write the first book. As I was about to decline the offer, I received three letters. It was a Friday morning, and I spent most of it in the back shed of our restaurant sobbing.

The first was from a woman who lives in Western Australia. She wrote that after reading *Secrets of the Red Lantern*, she felt an incredible sense of loss and guilt. Loss, because before reading the book, she never knew what it meant to be an immigrant, refugee or boat person. And guilt, because she had mistreated the immigrant kids at school. Now, twenty years after leaving school, she needed to make amends.

The second letter was from a couple living in Woollahra, in Sydney. They celebrated the wife's birthday at Red Lantern and bought a copy of the book. They wrote that some onions must have got caught in the pages, as they both sat up all night reading and crying tears of applause. They said that although it was a Vietnamese migrant story it was also their story, and their story had never been told in such a way before. They had fled Nazi Germany.

The third letter left me speechless. A man from Cecil Park, in Sydney's outer west, wrote eight pages. He too was a victim of child abuse, and after reading the book he was inspired and determined to return to his home country to try to find the reasons for his father's behaviour. He wrote that he was inspired to go back to Ireland to try to find answers, and compassion and peace.

The next thing to do was to ask my parents, and my nervousness and fear returned. When I told my father about the second book, his answer threw me. 'How can you say no? You cannot. There are so many people out there with stories to tell but no voice to tell it. You have been given this opportunity and you must say yes.'

So with the blessing of my parents and the powerful words of others, I have agreed to write a second book, another *Secret*...

Pauline Nguyen wrote *Secrets of the Red Lantern* with her brother Luke Nguyen and her partner, Mark Jensen, with whom she owns Red Lantern in Surry Hills, Sydney. Red Lantern has won many awards and in 2009 was named one of the *Sydney Morning Herald*'s most environmentally sustainable restaurants. Pauline graduated in communications from the University of Technology, Sydney.

FICTION

MILK

EDWINA PRESTON

IN the beginning, my mother took the form of a book. A 220-page paperback, with a cream and orange jacket, and the imprint of a small milk-bottle-shaped bird. *A Narrow Street*, boasted the sans-serif title that stood in for her eyes, and at the bottom, Elliot Paul. Or was it John Steinbeck and *Cannery Row*? I'm not sure. Mother came in many guises in those first weeks. And though the peculiarly shaped bird may just have been a black smudge against a less-black background (I had no depth of vision), I'm sure about the colour scheme itself; it did not change.

For some time I believed, as I gazed upwards chortling milk, that the book that was my mother radiated a mass of golden wavy hair. That it ate packets and packets of a certain digestive biscuit covered in chocolate. That it heaved and shifted and made low rumbling sounds, and sometimes sharp high barks. I was not unhappy: a steady stream of milk was being delivered. I was not cold; I was not hungry. I could rely on my mother to be constant in shape and unchanging in complexion. And then one day, she yawned and the book dropped from her face and everything changed.

I broke off from the breast in terror. For a start, Mother was enormous. Secondly, she was not orange and cream and black, and there was nothing perpendicular about her at all; she was white and gold and lacked precise edges. She had a large open mouth that was studded with small, terrifyingly precise teeth. And her face was made up of shifting contours and three-dimensional protuberances that broke and came back together again, expanded and contracted and merged, and provided no point of purchase at all.

This strangeness of Mother's was not a short-lived phenomenon. She never did congeal into a reliable shape. And so I have always seen her like this: a vast blonde topography I cannot comprehend the dips and falls of, woods I cannot see on account of all the trees.

MY SISTER, ORNELLA, does not believe any of this. She thinks it impossible to recall so early an experience. I do not tell her how much further my memories go. I do not tell her of my terrifying dream, the one in which I am inside a boa constrictor. I do not tell her how heavy

I am in this dream, nor of the terrible pressure, which comes and goes, getting more just when it seems to be getting less. When it is there, it is unbearable. It sucks and squeezes. It displaces space and air. It is a kind of airlock or vacuum, all gravity and no gravity, everything and nothing. And just when I can no longer endure it, suddenly it is gone; I am infinitesimally small and terrifyingly weightless. I am a tiny speck of dust on the head of a pin. There is nothing to contain me and nothing to stop me from floating away. And this lightness is far worse, far more frightening, than the weight that precedes it. It is the feeling of the beginning and the end. I would like to but I do not ask my sister what it means.

I must forget these things. Forgetting is useful. Forgetting is a clever neurological ruse. It pays for the brain to be forgetful.

Ornella has made a virtue of forgetfulness, but even she is not entirely devoid of memories. She once confided the recollection of a particular pair of red baby stockings, and how, with every yank of gusset over nappy, she was propelled two inches across the floor on her backside. From then, she says, there is nothing in her 'memory bank' until a grade-three skipping contest. Before she knew it, she was an adult. In between, I imagine large expanses of grey bitumen, tossed with greaseproof paper and orange peel.

I don't need to rely on memory alone regarding my mother and her books. There is evidence to back me up. Mr Knox, the bookseller, attests to the box of Penguins – he gave them to Mother; they were damaged or in some other way imperfect so he could not sell them, and she was so anxious, he said, that her brain would fall away.

The only memory I am prepared to concede is the one relating to Mother's vastness.

'Owen,' Ornella says, shaking her head, 'our mother was not a large woman. She was petite. *Pe-teet*.' When she says this, she brings together her thumb and forefinger, as though pinching out a lateral from a tomato plant.

The discrepancy must be due to childhood perception, for the one photograph I have of Mother at this time bears Ornella's claim out – she is not large at all; she is small and compact and wears a yellow terry-towelling dress with green buttons. I am in a carrycot made of woven blue and white plastic. Mother is not holding the carrycot; she is holding an orange leather suitcase. She is not going anywhere – not yet – and the suitcase, I now know, is full of manuscripts. Her legs are like a postman's: solid with muscle. And her hands are little paws; the fingers do not properly wrap around the handle. They are dough-maker's fingers. Pizza chef's fingers.

Shortly after this photograph was taken, I became Mother's enemy, for I had discovered my hands and fingers and feet. I remember the hard little shriek that came up from her throat. I can see her palm rise in the air as if to strike. I was a four-legged beast that clambered onto couches where books lay and ripped out frontispieces. I demolished the contents of the lower bookcase shelves, making no distinction between first and later editions. I chewed the corners of Thackeray till they were moist and soft. I flicked through pages – tick-tick-tick, with little hooklike thumbs – and then crushed them under my fist. When Mother settled down to a quiet afternoon read, the only thing I wanted to do was eat her book.

MY EARLY PREOCCUPATION with eating, and the iron constitution that allowed me to digest, unharmed, seventeen pages of Balzac's *Cousin Bette* (a book that goes on far too long, in my opinion), was a source of great pride to Papa. He boasted of how, prising open my teeth one day, he'd dislodged a shilling-sized lump of Balzac, a three-inch-long piece of hessian and a two-pound note. The laundry powders were removed from the linoleum floor, and one-penny pieces were attended to with great vigilance.

I was unstoppable, ate everything in my path; it is surprising I was not enormous. And therein lay Papa's second source of pride: for while I was no giant, the charts all clearly showed me to be 'above average'. This was an evolutionary triumph of the highest degree, because Papa was small, as all the men in his family had been. Not notionally small like Mother – really small. Barely five foot. He had expected the tiniest of offspring. He kept my first suit of clothes as testament to his triumph, but they do not strike me as big, only as very soiled and very little worth preserving. They are in a trunk with his father's war medals and a miniature silver-plated soup ladle Papa received for charitable services undertaken under the auspices of St Vincent de Paul. May 1961, says the soup ladle: the month in which he fed a record number of homeless and impecunious men in the electorate of Cleborne.

I have memories, too, of the kitchen where this charitable feat was executed. These memories are bigger and more boisterous than the ones relating to my mother. They feature what seem to be vats of capers and anchovies (though surely they are only large tins), slabs of bacon and orange freckled hocks that resemble orthopaedic shoes.

There I sit, amid the ruins of a record-winning stew, pinning currant eyes to a potato.

Mother is there too, somewhere, with a book hidden behind a menu, and a tray of liqueur plums hidden behind that. She would've had a bag by her side that contained items necessary to me: a cup with a lipped lid, an overnight diaper and a pair of pyjamas. Let no one say she did not care for me. She did her best to ensure the world's requirements regarding offspring were met.

But see, I have returned to Mother again — that is how large she looms — whereas, really, in the kitchen she was a marginal presence. The only true personality in the kitchen was the kitchen itself. Sometimes secret and cavernous, sometimes airy, sometimes panoramic, it was a marvel larger and more alive than either parent, and equal in wonder and illumination only to the library in the house of Mr Parish (though Ornella, I know, loathes this room, considers it a study in torture). In place of Mr Parish's books and paintings, the kitchen had coils of sausage and knots of garlic, colanders and tongs and spoons with holes in their handles. I have never been able to decide which of the two environments, the library or the kitchen, was the more exotic. But ultimately, the kitchen smelt better; I never liked the odour of shellac and turps I associate with the rooms of Mr Parish's house.

My earliest memory of the kitchen does not involve Mother or Papa, or any activities they might have been engaged in. My earliest memory involves a space between activities and people. A safe space, where puddles formed and dried and formed again and a curious cottony mould grew on the lino. I remember crawling into that space – I could not yet walk – to retrieve a morsel of food. I seemed to remain there for hours. There was warm enamelled steel on one side of me, whirring gently, and cold steel on the other, still. And I remember how secret and safe I felt, pressed in between those two contraptions, like a cat under a house in a storm. I swallowed the morsel that had lured me in, and discovered that the curious cottony mould could be unrolled from the lino and made into a ball, and wedged between the steel contraption and the floor. It was then, while so occupied, prodding and wedging and blowing my hot dark breath against the metal, that I felt one of the contraptions begin to move. A pair of white clattering shoes appeared at its far side, and the cold contraption began to roll towards the warm one on its casters. I felt a slow clamping pressure against my stomach and my ribs, a great squeezing that grew more and more intense. My chest compressed and my shoulders jammed up to make space, but there was no longer any room, and I suppose I must have made an oozing sound or a squealing sound, for suddenly the steel lurched violently away on the cold side, and the loud white shoes were upon me, and I was lifted into arms that smelled of oranges and cologne. And the bearer of the arms rocked me and pressed me into her neck and said, 'My God! Oh, my God!' – and that was Ornella. My first memory of Ornella. She was only just an adult, but she held me with more ferocity than my mother ever had. I loved her immediately. It was a spellbound kind of love: distinct and fully realised and entirely trusting.

Yes, I know, it's impossible, but the only talent I have ever truly had is this capacity for preternatural recall.

Edwina Preston is the author of *Not Just A Suburban Boy* (Duffy & Snellgrove, 2002), a biography of the Melbourne artist Howard Arkley. She has contributed articles and reviews to major Australian newspapers. This is an extract from her novel-in-progress, *The Pepper Experiment*.

Scenes from life with father

Failing the sniff test

Sydney Smith

PICTURE a winter's evening in a kitchen in Wellington, New Zealand. An Antarctic wind stalks the house, rattling windows in its quest for a way in. I'm sitting at the table while my meal goes cold. My brothers bolted theirs down an hour ago and escaped to their bedrooms. Mother says, 'You'll sit there until you've eaten every bit.' A bare light bulb burns harshly yellow overhead and picks up a thick sheen on the sliced ox tongue and the boiled potatoes. Only the spinach, with its strong green colouring, withstood the luminous onslaught. I like spinach, especially when it's tossed with salt. I ate that first and have stared disconsolately ever since at the rest. I have tried the potato: it's as hard and unappetising as soap.

More particularly, though, I don't want to eat the tongue. It's a shocking thing to tell a child to eat, an obscene thing. I saw the tongue when my father brought it home, unwrapped it and set it on a plate. It was blue-grey then, and grossly pimpled. All innocently, ghoulishly curious, I watched Mother boil and peel it, thinking it was for my parents to eat, not us kids. I know there's adult food and child food, even if nobody has said so.

When we sat down to the meal, I said, 'Why do we have to eat tongue?' Mother said, 'Your father wants it.' The king's wish is our command. His penchant for weird animal body parts – brains, liver and kidney, tripe, sweetbreads – is mysterious enough. Even more mysterious is that he eats his tea at the wrong time of day. Mother puts a lid over his share and slides it into the oven. Early the next morning, while the rest of us are sleeping, he will heat and consume it with a cup of coffee for breakfast. I sometimes see his plate in the morning, ringed with congealed gravy.

When I think of my father, I always think of revolting food.

IT'S AN OPEN secret in my family that my father can't smell. Mother told me he lost his sense of smell when he was nine, in 1932, during a botched operation to remove his tonsils and adenoids. The procedure took place on the kitchen table, she said. (I picture a white-coated man carving a roast. Do humans eat the tonsils and adenoids of animals?) But though we all know, none of us talks about it in front of him. We know it's a banned subject, even though we haven't been told it is.

The operation had one interesting result: my father can sniff up one nostril at a time. Steven and I hang around him when he comes home from work, just for the chance to see him do it. He has a gigantic schnozz, a source of fascination in its own right. It makes our day when he brusquely snorts through his right nostril, then his left. Steven says he'll hold a piece of paper under the nose one day and see if that gets sucked up.

My father's inability to smell has one very bad effect. He cooks the Sunday tea most of the year, except during the lawn-bowls season in summer; he cooks his favourite dishes at other times during the week. He prepares pea and ham soup with too much salt, cooks curried sausages so hot we turn red and breathe fire. He liberally coats his share of meals with salt and white pepper. His meals are a test of endurance my older brothers boast of to their friends. I laugh when I hear them challenge a boy to eat a curried sausage, and laugh harder when the boy cringes; but I wonder too why food has to be a trial in our house.

My father's inability to smell has one brief, noteworthy consequence.

WHEN MOTHER WAS growing up, it was the policy in white New Zealand to 'assimilate' the Maori. Mother was sent by her parents to a boarding school devoted to inducting Maori girls into the *pakeha* way of life. The girls were strapped if they were caught speaking Maori. They were given elocution lessons to banish their Maori accent. And they were taught to reject Maori food and cook pakeha meals.

Maori food is fish and shellfish; eels; birds and their eggs; *kuumara*, the Maori sweet potato; and *puha*, green leafy vegetables. Plenty of non-Maori people eat these foods; but to me, Maori food is not just the victuals themselves: it is the way they are cooked, and an indefinable something else which I recognise as soon as I walk into the home of one of my mother's older siblings and which is absent from ours, something about the smell of the air, the buckets of blue-black mussels, of pipis still coated in sand; the sight of degloved eels draped over the lip of the kitchen sink, how wincingly raw they look.

The Maori traditionally cooked their food in an earthen oven, or *haangii*: it was dug out of the ground by the men and lined with river stones which had been heated in a fire until they were red-hot; the birds, fish and kuumara, prepared by the women and wrapped in leaves, were placed inside the oven, covered with leaves and the earth shovelled on top. The food was left for the afternoon, then dug up.

Haangii food is cooked at all the big family events Mother takes me to, especially funerals, the most important occasions in Maori life. I sit among my aunties and uncles and their children and our cousins once or twice or thrice removed, all of us at long trestle tables, and watch in puzzled revulsion as they tuck in with gusto, talking and laughing all the while. I can't stand the acrid whiff of it and won't eat, despite Mother's urgings. I embarrass her on these occasions: among the Maori, as with all ethnic groups, it is rude to reject a host's food. I can't help myself, though. It looks unprepossessing, smells horrible, and tastes of smoke and dirt. (I have discovered since that haangii food, when properly cooked, doesn't smell or taste like the food cooked in my uncles' haangii. They used dampened sacking to steam the bundles of meat and vegetables, which is the modern custom. I think they neglected to clean the sacks, which were impregnated with the soil of the potatoes they had held before they were collected from the greengrocer.)

At home, with no special occasion to celebrate, my aunties boil pots of pumpkin. I often see them eat a big pile of the vegetable with no other flavouring than a dab of butter, and no other foods to vary the feast. They eat with inexplicable zest. I like pumpkin well enough, in small quantities. I can't understand why they eat piles of it this way. It's as if food to them is something different to what it is to me, perhaps an assertion of their racial identity, perhaps a happy reminder of childhood, perhaps even a sign of afflu-ence in contrast to former years when food wasn't always plentiful.

In our house, Mother cooks the kind of food my pakeha father approves of: roasted meat, stews, fried food and a dish I loathe, in which a magenta-coloured wedge of corned beef is put in a large pot on the stove, to which is added water, then carrots, potatoes and peas. The whole lot is simmered until everything tastes like the salted meat.

Mother cooks Maori food only once, and does it with an openness that belies the secretiveness of the enterprise. One of her cousins delivers a brace of mutton-birds, wrapped in newspaper, which she hides at the back of the fridge behind a gaggle of jars. Mutton-birds thrive on Stewart Island, the tiny anchor at the stern of the canoe-shaped South Island. Being a sub-Antarctic creature, they have a dense layer of fat between the skin and the flesh. The smell when they are cooking is eye-wateringly pungent, and the result so greasy that Mother has to don a plastic apron before she sits down to this delicacy. She spends a good hour over the business, and at the end her mouth, cheeks, chin, hands, forearms and apron glisten with fat. She also vibrates with satisfaction, the kind of satisfaction I routinely see in her sisters when they eat a dish of boiled pumpkin. For days afterwards, she reeks of mutton-bird.

She prepares the treat one Saturday. My father is always out on Satur-days. She washes and puts away the cooking pot and the plate she has eaten from before he gets home. She doesn't need to worry about the stench of mutton-bird, of Maori food, which has invaded every nook and cranny of the house.

I wonder why this is the only time she takes advantage of my father's disability. Perhaps it's that every time she visits her sisters, she gets to eat Maori food – except that I never see her enjoy that the way she enjoys the mutton-birds.

DURING THE LAWN bowls off-season, my father bestows on us the privilege of a Sunday drive. This is an event I dread. As soon as we get into the car, he points a warning finger and growls, 'Don't say a word. I have to concentrate.' Mother doesn't think the rule applies to her and chatters the whole way. Father interrupts her monologue to say, 'For Chrissake, will you shut your trap! I'm driving. Do you want me to hit a lamppost?' His voice is as clenched as his fists on the steering wheel. Nothing he can say will make her be quiet. In fact, she seems to enjoy upsetting him.

In this way, we trundle around the bays, which are full of choppy grey water. Out of fear of hitting a lamppost, my father drives well below the speed limit. Other cars pass us. So do cyclists. A granny with a walking stick hobbles by. My brothers dismally watch these overtakers and shoot our father accusing glances. Nothing is said but even I, the girl, know that his driving lacks manly vigour.

On the homeward stretch of what to my father has been a reasonably successful outing, despite Mother, he stops at the only milk bar in Wellington that sells jaffa ice-cream. He digs his hand into his trouser pocket, pulls out a fistful of change and gives my brother Lynn some money. A few minutes later, he returns with a bouquet of orange ice-creams in cones. He hands them through the driver's side window. While my brothers and parents tuck in, I hesitate. My father glances at me in the rear-view mirror. 'Eat up before it melts,' he says. It is already melting in rivulets, possibly warmed by the fever of disgust coursing through my veins. I poke out my tongue gingerly and touch the orange ball flecked with brown chocolate chips. It looks like something best flushed down the toilet. My stomach is already turning queasy somersaults. I can't tell him, because I know he won't listen. I feel my eyes bulge with the fear of what my stomach is about to do. My father starts the car and snail-crawls away from the kerb. Shortly after, he pulls over and opens the back door. Nothing, not even his baffled disapproval, can stop the vomit I hurl into the gutter.

It doesn't occur to anyone to excuse me from jaffa duty, least of all my father, who must assume each Sunday that, this time, at last, I will appreciate the treat.

MY FATHER BUYS the fruit and vegetables each week. Although we have moved from suburb to suburb in Wellington, each Saturday he visits the same greengrocer, a man who understands what he wants. To my father, a bargain is the uppermost consideration when buying anything, especially food. At the end of the trading week, the greengrocer sets aside everything that is spotted, bruised, spongy, wounded, and offers it to my father at half price. He snaps it up.

While we were young, he bought the potatoes each week. But after my brothers left home, with more money at his disposal, he seized the chance to save by periodically buying a sack of potatoes. It is my job to peel and prepare them for roasting. I enjoy cutting the string that sews the lips of a new sack together, because it means that for a week or two our potatoes will be delicious. Nothing can compare to firm creamy flesh coated in fat and baked until crisp. But as the weeks pass and the sack slowly empties, the potatoes become spongy, sprout, grow spots of blue fungus and take on a bitter flavour. I know it's useless to tell him that it isn't a bargain when the potatoes are inedible by the time we're halfway through the sack, that it's excessive to buy a sack when there's only him, Mother and me to feed. My father can't hear anyone else's opinion.

Mealtime is a species of torture, minus the screaming. I eat my peas, beans and carrots, because they are the least offensive items on my plate. I linger over the rest, talking myself into believing they aren't as bad as they look. I feel my father watching me for signs of rejection, his expression suspended between anger and hurt. Whether he chooses to growl or hang his head in pain, the outcome for me will be the same: his happiness depends on my acceptance of his food.

LYNN MOVED TO Australia for work, and when he comes back he has new recipes to dazzle us with. He discovers that our father has taken over all cooking duties and takes appropriate action. For the week of his stay, he ensures that he, not my father, cooks the evening meal. He introduces us to stir-fried vegetables with a pinch of curry to heighten the flavour. Not only that (and that is dramatic enough to me), he buys the vegetables each day, fresh from the greengrocer's. Not only *that*, he buys the best, free of spots, bruises, fungus, tears and holes in the leaves. It is heaven to eat green beans that taste like beans, cauliflower given colour and speckles by the curry powder, and to try new vegetables, broccoli, red pepper, all feeling deliciously firm-soft in the mouth.

A week later, he flies back to Melbourne and our household returns to my father's regime. With changes. My father inexplicably decides to cook stir-fried vegetables – Lynn has told him how. His reason is inexplicable because my father learned how to cook five or six dishes in his youth, while training to be a cook at the People's Palace, and ever since has stuck to them like grim death. But he announces to Mother and me that he is cooking stir-fry, and that is what he does five nights a week for the next year. On the first night, he closely follows my brother's instructions and measurements. After that, he goes his own way. He has no sense of smell to guide him. He doesn't know the difference between a pinch of curry and a tablespoon. Well, he understands the difference in quantity, but not the effect it has on Mother and me. Unused to sautéing vegetables, he cooks them to mush.

Also – and this comes out in amused asides to Mother – he uses cheap, damaged vegetables. Though he holds his eldest son in awe for living overseas, for working in an office, for having spent a semester at university (he and his pub mates passed the cap around to buy Lynn a leather briefcase engraved with his initials at the start of the term), he views him with scorn for being sucker enough to buy vegetables at full price.

He gives up the stir-fry tyranny after he and Mother visit Lynn in Australia. He comes home with a crockpot and makes Hungarian meatballs, to Lynn's recipe, every Saturday for unnumbered years afterwards. Nothing can deter him from his culinary course, not complaints from Steven when he visits on Sunday ('Hungarian meatballs *again*'), not uneaten meatballs idling in a puddle of yellow grease and which he has to throw out. He has no concept of the false economy, that food which isn't eaten is dearer than food which is, no matter how much more it cost to buy.

AS I TURN from adolescent to adult, I think of my father as a nub of survival, every thought bent solely on the task of living from day to day, and oblivious to others. I don't think of him all that much, because I have other problems to occupy my mind. For example, people can see my thoughts through the transparent cover of my forehead. Though I hurry along the street, my head ducked protectively low, I know it's useless, that everything I think and feel is on display for all to see and condemn. So I stop leaving the house. I stay in my bedroom day and night, coming out only to use the toilet and get something to eat. I stop washing because of the cameras behind the bathroom mirrors which record my movements.

It comes as a surprise when I realise my father knows something is wrong. I thought nobody had noticed. He rings me every morning during smoko. For the first few months, I answer when the phone rings, because that is what people do. Nobody else is home to do it. He asks me what I want for tea. He asks me to do this task or that. When I cry, he quickly backs down – 'I'll do it tonight,' he says. But after these months have passed, the sound of the phone is a nagging pressure, an inconsolable child who won't stop its frightened crying. He's scared I will kill myself and he wants proof each day that I'm still alive. Sometimes, I don't answer. I count the rings: eighty-two, eighty-three, eighty-four. The phone falls silent. Has he got the message? No, it starts again a few minutes later. He must have been called away for something. That night, when he comes home, he interrogates me about why I didn't answer his call. I don't know how to tell him how much I hate the sound of the telephone.

The butcher's shop sends him out twice a week on delivery rounds. He drops by our place to leave plastic sacks of frozen chips, parcels of meat and groceries. At night, as soon as he gets in from work, he scrapes animal fat into a big saucepan, lights the gas ring under it and, when it begins to smoke, pours in frozen chips. As the ice hits the hot fat, a battle ensues, with roars and crackling explosions. From my room, I hear him shake the pan to break up the clumps of potato. He scoops the cooked chips into an enamel dish and calls for me to come and eat. At first I do. I like chips. But he cooks chips day after day, week after week, month after month, and by the time two years have passed, I have stopped eating them and, behind his back, throw them in the bin. So as not to hurt his feelings, I hide them under rubbish and, since that never seems enough, I crumple newspaper and shove that on top. I try to tell him I can't eat chips every day. By this time I am seeing a therapist and have started to assert myself in small, timid ways. I know it's useless but still I try. He wants to help me, but everything he does is pressed in and squeezed tight and made rigid by his belief that life is a grim struggle for survival. He doesn't listen to me, not even when, out of desperation, I invent a story of the doctor putting me on a no-fat diet.

In the end, I stop the chip regime by leaving the country. I know in hindsight that nothing less could have stopped him.

I COME TO live in Melbourne when I'm twenty-five. Although I'm still sick, I know that if I want to get well I have to get far away from my old existence. The day I told my father I was leaving home, he asked, 'Why?' I said I wanted to live my own life. Baffled, offended, he asked, 'Why?' As we all knew in my family, I had been earmarked to take care of my parents until they died. I had entertained garish fantasies in which I withered to a grey stick and died the day they did. Perhaps he had had a similar fantasy – though if he had, its tone would have been very different. Whereas mine had the bitter tang of a curse, his would have been like the fulfilment of an anxiously held wish: to protect his daughter forever from the dangers and depredations of the world. My decision to leave his protection was not simply an act of rebellion but of insult, a rejection of his kind of love. During the month after I told him I was leaving, while I prepared for flight, he didn't speak to me.

TO MY FATHER, food was just something you put in your stomach to stop the hunger pangs. It had no joy, no pleasure, no opulence. It was function. This attitude sprang from his anosmia, his inability to smell. Taste, which we experience through the tongue, comprises a third of the sensations we enjoy when we eat and drink; we taste sweet and salty, sour and bitter. The rest, which is called flavour, is supplied by smell. My father had lost on that long-ago operating table most of his ability to enjoy food. He had lost any sense of the richness and complexity, the experience, of eating.

Nevertheless, anosmia can't explain his peculiar rigidity. I wish I could understand him better. But he didn't talk about himself, didn't speak of his past, didn't mention his parents, who were dead before I was born. I will never know what forces of austerity shaped him into the man I grew up with. All I have is the petrified residue of his habits in which he is captured, a fly in amber.

The last time my father cooked for me, I was living in a shared house; he and Mother had come over for a holiday and to attend the Melbourne Cup. They were staying with Lynn; they didn't want to see where I lived, as it would remind them too sharply of what they had lost. On this night, my brother was out and would not be back until late, so the cooking duty fell to my father. He boiled a lump of pickled pork, peeled potatoes, chopped cauliflower and cabbage. It had been three years since I had eaten one of his meals, so the dinner is printed indelibly on my mind. It was served on white plates, and the only colour to be seen was the pale mucous pink of the pork, the washed-out green of the cabbage and the black spots on the potatoes. I stared in disbelief. By this time, I was used to other kinds of food and had a whole new vocabulary to go with them: risotto, creamy hummus, beef strips threaded onto bamboo sticks and reclining in the aromatic ooze of peanut sauce. My father's meal was a shocking reminder of another life, one of hunger despite a full stomach. He saw my reaction and growled at me to eat. I managed half a potato out of duty and pushed my plate away. Cruel as I knew I was being, I couldn't hide my revulsion. I couldn't do it to please or placate him, and I couldn't do it for love.

Sydney Smith is a past winner of *The Age* Short Story Competition. Another of her memoirs was published in *Griffith REVIEW 22: MoneySexPower* and her story 'Flame Red' appeared in *Griffith REVIEW 26: Stories for Today*. She is the director of the Victorian Mentoring Service for Writers.

Smallgoods

The first taste of mackerel

Eva Lomski

TELL me a story, my daughter says, and even though I don't feel like telling stories tonight, I still tell her about the best way to eat smoked mackerel.

First, I say, you have to travel back in time to 1974, when I was exactly your age. This is too hard for her to imagine, so I say: Remember the *Brady Bunch* DVD?

Her eyes gleam and she sings *Here's a story / Of a lovely lady*, and sits up in bed so quickly Tilly Tiger threatens to knock off the rest of her snout by falling to the floor. She relishes the opportunity to clamber out again and says: If we're doing *The Brady Bunch*, then can I be Marcia?

Okay, I say, you can be Marcia, but you live in Adelaide and you've just finished Saturday morning ballet class at the Gwen Mackay School of Ballet...

Am I on points? she asks, back in bed with Tilly.

Yes, you can be en pointe.

Cool.

...in this big old building that looks like a monastery, with pentagon façades and doorway arches and redbrick detailing, right next door to Parliament House. Now this was before it became the Constitutional Museum...

My daughter jabs me in the ribs.

…and it had lots of creepy staircases and dark hallways, just like Hogwarts.

She smiles the same way she does when I finally agree to play shoe shops after hours of negotiation.

When your mother picks you up after ballet, you run, still in leotards, across the road with Zosia, the girl with the lazy eye I told you about…

The one they laughed at?

…that's the one – to the railway station and continue running and laughing down its entrance ramp, running so fast it feels like falling, down and down like Alice in Wonderland. At the bottom there's a kiosk that looks like it's been there a hundred years but it sells caramel Crazy Mazes and you and Zosia buy one each because you both love the way it stretches when you hold it in your teeth and pull.

Is the railway station like in Harry Potter, too?

Not quite, I say, but close. It has a marble hall two storeys high and telephone booths set into the wall, just like you see in old movies with Marlene Dietrich in them, all heavy black handsets and buttons. This was before whoever it was sold out and part of it got converted into a casino…

She reaches out and jabs me again. The mackerel?

Then you and Zosia and your mothers head back up to the car park and you wave goodbye and yell: See you next week! And Zosia smiles the biggest smile ever and her mother looks as if she might cry. You sit in the back seat of the grey Datsun 1000, which looks like a slater on wheels, and your mother talks to herself as she drives down North Terrace; past the trees that would remind you of Paris if you'd ever been to Paris; right into King William Street; and through the city centre to Victoria Square, where the police building has goldfish in a pond out the front. A duck lays her eggs there, and each year the police stop the traffic and escort the duck and her line of ducklings all the way down King William Street to the Torrens River.

That's so cute, my daughter says. When does the mackerel come into it?

Take a right into Grote Street and head up to the car park on the roof of the Central Market. Even before you go downstairs you know you're in the right place, because the smell of hot roasted nuts grabs hold of all the smelly bits up your nose…

Eew!

...and doesn't stop squeezing until you beg your mother to buy you a bag of nuts. She says no, reminding you of the time two years ago when you dropped the bag and an old lady slipped on the cashews and yelled so long they had to call an ambulance.

Was she okay?

I guess. I never knew. Anyway, you're downstairs now and your mother knows every stall in the market but you get mixed up because one fruit stall or delicatessen looks pretty much like any other. Also, the stalls are set higher, so the customers have to look up and it's even harder to see if you're that much closer to the ground.

Am I still in the leotard? my daughter asks. I won't go anywhere in just a leotard.

You put on a skirt in the car. Now it smells of sausages and vinegar and you come to a stall which has all its creamy cheeses arranged in a long line to the right of the counter. Your mother is greeted with a smile by the Ukrainian woman who wears a headscarf to hide her bald spot and always sneaks you a slice of ham to fatten you up, so she says.

Where's Ukrainium?

Ukraine's in Europe. *Dobrý de,* your mother says, because back then you knew what everybody was and you all spoke a few words of each other's language. Your mother ums and ahs and steps back and frowns, but then buys what she always buys: a kilo of the Neufchâtel cheese that feels like you're eating toothpaste but tastes like heaven; the smoked mackerel; Polish dill pickles – real pickles, none of these made-in-India pretenders you get in the supermarket; white crusty bread, thick and dense; sausages you need to boil called parówki, your father's favourite; three other types of sausage; double-smoked gypsy ham – real ham, not like this stuff today all covered in preservatives, as slick as pig spit...

Double-eew!

...and a bucket of sauerkraut. On the way home, the Datsun is so full of the smell of fresh bread and sausage that you think your stomach will eat itself. So you beg your mother to take a detour home down Port Road, to the cake shop run by women who make everything themselves with real cream.

Even though your mother complains there's no time to waste, you end up watching the clock in the shop for thirty-five minutes with your stomach rumbling while she stands chatting in Polish, and you're in agony trying to decide between the striped vanilla cake and coffee-cream triangles, the baked cheesecake and the pâczki, which are donuts with no holes covered in sugar, with jam in the middle. If you're really lucky, your mother will be in such a good mood by then that she'll buy you one of each.

At home in the kitchen, you help by switching on the kettle, getting out the teabags and turning on the lights, because even though it's early afternoon the carport blocks out all light to the kitchen window. Your mother slices and dices and puts out plates of food on a plastic tablecloth that is spotless but has been there for as long as you can remember. You pat Irena, the fat corgi, whom you named after your father's favourite singer, Irena Santor, because of her blonde hair, and trace the swirls on the olive green wallpaper with a finger until your mother tells you not to. The Black Madonna looks down from the wall over all of this and your mother sings to herself *Holy God, Holy Strength, Holy Immortal, Have Mercy on Us.*

Is this before or after grandpa died? my daughter asks.

It takes me a second to reply. Before.

I'm tired, she says.

You're too hungry to wait any longer, so your mother tells you to sit and you use your special fork with the pattern on the handle to pick out sausage and bits of mackerel and set it on your plate. You take some bread. Then this is what you do. First, you take a sip of sweet milky tea, to see if the temperature is right. Then you take a bite of mackerel, and the smoky-oily flavour hits your tongue and you feel the texture; it should be firm yet soft, and not too dry. Then you crunch into the white bread and try not to spray any crumbs, but eating bread without making crumbs is like smelling flowers without any scent. Next you bite the Polish dill pickle until the salt and garlic, smoke and bread mix together like old friends, and you realise you're eating the most perfect combination of food you could possibly hope for. You take another sip of tea, your stomach sighs, and you're the happiest you've been in ages because your ballet teacher praised your pirouettes, you feel good about asking Zosia to go for chocolate, your mother's singing in the kitchen, there's cake coming,

Irena is lying under your feet, and you are waiting for your father to come home from work and tell you a funny story about his day. Suddenly your mother says: Oh, God. Did I give Bronek his medication before we left?

What? says my daughter.

I shut my eyes for a moment, then continue.

And you turn to your mother and say: Mother dearest, thank you for the glorious fare. This is surely and truly one hundred per cent better than McDonald's.

You're making that up, says my daughter, but her heart's not in it, and she is yawning.

Now it's time for the sausage, and you slap Neufchâtel on bread (none of this butter business) and finish your tea. You decide on the triangle cake and wonder how on earth they manage to slice the cake thin enough to make stripes. But nothing, not even the triangle cake, tastes quite as perfect as that first mouthful of smoked mackerel.

When can I try some? my daughter says as her eyelids flutter and Tilly Tiger relaxes.

Dear God – not tonight, my baby, I whisper, because by then she's asleep, and a whisper is all I can manage.

Eva Lomski edited the anthology *20 People Write: Short Stories & Poems of Life, Love, Loss & Whimsy* (Sandybeach Centre, 2008). Some events in this autobiographical account have been fictionalised.

MEMOIR

My happy
Cold War summers

In memory of my mother

Nikola Gurovic

SINCE early childhood, I have had a certain perception of the Iron Curtain and the Cold War. Not because I showed particular interest in politics and the military instead of soccer, cowboys and Indians, but for the simple fact of geography. My maternal grandparents lived in the plains on the border between Tito's Yugoslavia, which steered away from Moscow's rule in 1948, the year before I was born, and Hungary. The latter, during my boyhood and youth, was a loyal member of the Warsaw Pact and more or less an obedient follower of the Soviet Union. That loyalty, I would learn years later, was extorted by Soviets after their tanks crushed the Hungarian uprising in 1956.

My mother was from a village where life appeared in slow motion, except for the hard work in the fields. About eight hundred households stretched in a geometrical shape, occupying the northern tip of Yugoslavia. There, the railway would end abruptly, and my laidback summer holidays would start gently.

The railway station, a simple white two-room building with a stork's nest on the chimney, was almost a kilometre from the first houses and orchards, and at least two kilometres from my grandparents' home. The name Djala appeared in huge black letters in the Cyrillic and Latin alphabets above the doors for departing and arriving passengers.

MY UNCLE LÁSZLÓ, sitting on the cart drawn by a horse, would wait in the shadow of a huge acacia tree for an early afternoon train bringing my mother, sister and me from the southern region of Yugoslavia, where my family settled when my father decided to leave the military and start a career as an accountant. The trip took twenty-four hours.

I admired László, my mother's youngest brother, because he knew everything about horses, dogs and hunting, and he played soccer for a local team appropriately named Border Guard. I hated the first moment of our encounters, when he would give me his strong bear hug and an awkward kiss which made my cheeks prickle, as he only shaved once a week, usually on Sundays before the soccer match. 'You are much taller than last summer,

PREVIOUS PAGE: *Young chefs.* Photographer: Graham Monro/gmphotographics.com.au from the *Sydney Morning Herald* Shoot the Chef! 2009 photographic competition.

Mickey,' he would say, smiling, before taking from his shirt pocket a cheap, strong cigarette made of dark tobacco.

With hasty movements, he would arrange our luggage on the back of the cart and direct me next to him on the timber bench at the front. After handing me the reins and giving orders to a quiet chestnut mare, he would start the usual conversation with my mother, informing her about things he couldn't write about in his rare letters.

As we took off, the cart would leave a cloud of mustard-coloured dust behind us. On my right side, the watchtowers of the Yugoslav border guards punctured the flat horizon of sky and golden crops of grain and green fields of corn.

Over that strictly observed borderline, between the microcosms of my childhood summers and the big unknown on the other side, there was a long invisible strip of no man's land, and there were young Hungarian soldiers with their binoculars and machine-gun nests atop their timber watchtowers.

'They have camouflaged tranches, bunkers, artillery positions, tanks and arsenals. But we are not scared, are we, Mickey?' my uncle asked, stretching his right arm towards the Hungarian territory. Sensing my hesitation, he continued, more to himself: 'In the worst-case scenario, the Americans will help us. They can't afford one more domino falling into a Soviet lap.'

At age eight, I could only begin to imagine what would become of my summer-holiday oasis if the Kremlin ordered an invasion of Yugoslavia to end its status as a buffer zone between East and West, NATO and the Warsaw Pact.

Many summers have passed since those days, but I still vividly remember how religiously my grandfather István would listen to the Voice of America on his radio set that featured a huge compass-like display with a half-blue, half-red needle pointing to the various stations and a loudspeaker hidden behind cream-coloured rough canvas. The Kosmaj radio set (named for the mountain in Serbia famous for the intensive guerrilla operations of Tito's partisans during World War II) was a proud product of the Yugoslav electronics industry, and the most valuable item in the humble household of my grandparents. No wonder Grandpa would turn off the radio at the tiniest sign of a summer storm. To be absolutely sure, he would even disconnect the radio from the power socket.

'The lamps will burn out if lightning strikes,' Grandpa would explain in the voice of a surgeon ready to perform a complicated operation. Mind you, he was a cobbler. He just knew that the radio provided him with the most important insight into Cold War politics.

Although he didn't have any idea about propaganda and biased reporting, he would regularly tune in to Kossuth Radio Budapest and Radio Belgrade to be informed about the most recent directives of the Hungarian and Yugoslav leaders.

'Kossuth Radio plays my favourite Hungarian folk songs,' he would say after being glued to the radio for more than an hour, during which the only music was a minute or so of a signature tune borrowed from some traditional Hungarian song. He considered it his patriotic duty to listen to Radio Belgrade, so he didn't bother to offer any explanation for listening to it.

MY GRANDPARENTS' HOUSE, an old structure of washed-out yellow brick and mud, with two white-framed windows facing the street, mulberry trees and brick pedestrian path, was close to an abandoned cemetery. The wide wooden gate led to the front yard, a path lined with roses and a water pump. In the backyard there was a barn and a spacious vegetable garden with two small sour-cherry trees, a big apricot tree and a chestnut tree.

As we drew closer to my grandparents' place, the mare would instinctively hasten her pace. From a distance I would recognise a tall, slim figure. My grandmother Ethel, with her white hair, tortoiseshell glasses and worn cotton dress, was waiting at the doorstep to embrace me, her eldest grandchild who, as she often solemnly declared, would go a long way and never have to work on the land and depend on nature's kindness or the mercy of the communist bureaucracy.

Grandpa István was a corporal of the Austro-Hungarian military in his youth, and never gave up his commanding attitude. I could not say which was shorter, his stature or his fuse. He was never there to greet me upon my arrival, always inventing the most unlikely excuses to avoid that first encounter when he should show some care and tenderness.

'Did you get good marks in school?' he would ask, coming out from nowhere to take the head at the dinner table. 'He got excellent marks,' my mother would say, coming to my rescue. But Grandpa didn't seem impressed at all. 'From tomorrow, you'll fetch water twice a day; as for the other chores, ask Uncle László and Grandma. And keep away from Jana's grandsons,' he warned, knowing very well that I would do the opposite.

Our neighbour's grandsons, Aleksandar and Petar, my cousin Andor and I would spend long worriless summer hours playing soccer or marbles, making model planes, flying kites, hunting birds with slingshots or bows and arrows with no success, or taking long rides on battered bikes over the vast plains of Vojvodina, the northern province of Serbia.

Evenings were reserved for politics after a hard day in the fields. The neighbours, mainly men, would strategically position their miniature three-legged wooden chairs in front of their houses, smoke and discuss the latest news bulletins from Radio Belgrade, Kossuth Radio Budapest, Radio Moscow, BBC London, Voice of America…I would listen, half asleep, without understanding much except that the Russians and the Americans had terrifying weapons at their disposal, weapons that could obliterate the entire planet in no time. The smoke from their cigarettes and pipes would assault my lungs, but it was an excellent mosquito repellent. Like everything else in the fertile fields of Vojvodina, mosquitoes grew bigger and more aggressive than normal.

THE WATERSHED EVENT would occur in the second or third week of July. The wheat was harvested from the faraway fields. It would be bundled and stocked in the backyard, waiting for the thresher. The owner of the only thresher in the village was the father-in-law of Uncle Gyula, my mother's middle brother. This meant easier negotiations about the threshing date.

The process required a huge amount of labour, skill, military-like logistics and perfect timing. Lajos, my mother's oldest brother, would come from Novi Sad to lead a team of a dozen neighbours and relatives. Early in the morning, as soon as the wooden threshing machine, powered by a steam tractor, started making an unbearable noise, my mother would join her brothers in standing on top of the mountain of wheat and pitching down the grain bundles into the threshing machine's bundle feeder.

My cousin Andor and I would run around as if we were the most important workers in the line. Early in the afternoon, when the job was close to
finished, Grandpa would count the bags of wheat and jot down numbers.
The bigger part of the harvest would be delivered to the local baker, Ivan, in
exchange for bread supplies during the year. The remaining grain was to be
sold at state-controlled (and usually very low) prices.

Ivan, a former wrestler, came to Djala from Croatia in his early forties.
His best friend was Hector, a well-trained hound. After the morning rush,
Ivan would shout: 'Hector, go get my paper!' A quarter of an hour later, the
four-legged courier would return with the latest edition of *Politika* between
his teeth. Ivan was proud of his dog and passionate about his trade. He baked
three-kilo cobs of bread with three-millimetre-thick brown shiny crusts.
Andor and I would often watch him knead the dough with his strong arms.
Coming from the city, where you could only buy a loaf of bread or a half, I
watched with interest how skilfully Ivan would cut his wheel-sized breads
into thirds, quarters or even smaller, depending on the customer.

WHILE OFFERING A small glass of mulberry or plum brandy to the
workers waiting their turn to wash their hands and faces, Grandpa would
whinge about another average season. Everyone was waiting for Aunt Terez
to appear and announce that lunch was ready. This was the moment Andor
and I had been waiting for since the early morning. The threshing-day lunch
was special, a festival to celebrate the harvest, the gathering of family and
good food.

With my grandma as an agile assistant, Aunt Terez would prepare
meals that could be served in the best restaurants of Belgrade or Budapest.
Homegrown vegetable soup with carrots, kohlrabi, cauliflower, potatoes,
cumin, pepper and sour cream were a natural start to one of the most significant lunches of the season. Fried chicken and pork cutlets with stewed green
beans or baby peas with a touch of dill would be the main course.

For years after the war, food was scarce in many regions of Yugoslavia,
but not in Vojvodina, which had some of the best soil in Europe. Hard-working
farmers were producing more than they needed. The communist system forced

them to sell the surplus at ridiculous prices. For that reason, Grandpa eventually gave up his notes and figures to get beer and white wine.

At the same time Aunt Terez, with a Napoleonic talent for commanding, would observe the field, making sure everyone got enough to eat and drink. At some point, passing by the table where Andor, my sister Anica, cousin Marika and I would sit, she would suggest we save room for the best, which was still to come. Cooking from the early morning, she still didn't have time to prepare apple or walnut strudel, or baked pancakes filled with ricotta-like cheese, sugar and cinnamon or apricot jam. Instead, a huge steaming bowl of fettuccine mixed with ground poppy seeds and fine sugar would land on the table, making even the meat lovers think twice.

For the few who didn't care much for poppy-seed fettuccini, ice-cold quinces and sour cherries preserved in sugar syrup would neutralise the heavy taste of fried chicken and pork. The long and rich culinary tradition of the plains involved much knowledge about the fine balance of various ingredients. Fertile soil and diligent work provided families with top-quality vegetables, fruits, meat and diary products. They were never treated fairly by the state. The only reward they could enjoy was the wonderful food they produced.

LATE IN AUGUST, Grandma would usually invite me to join her in her brother's vineyard: four lines of vine running down the slope, only a hundred and fifty metres from the border. The grapes were still several weeks from being ripe, but on three or four plants, early grapes were ready for consumption.

Light green in colour and the size of a pea, the grapes looked unspectacular – but the taste was so fresh, so delicate, unique. Sitting on the back veranda, overlooking the vineyard and the border, I would devour the first grapes of the season imagining that some crazy explorer brought the fruit to this part of the world directly from Shangri-la. The delicate harmony of tastes – the acidity of gooseberry, the strength of sugar, the smooth freshness of kiwi – made those grapes heavenly. Half a century later, I am still searching for the fruits of my childhood and longing for my summer oasis.

Sometimes in my dreams, I hear my grandmother speaking softly in

Hungarian: 'Wake up, my boy, the holidays are over, you have a long journey ahead.'

With heavy eyelids and a sleepy mind I would mix the buttons and holes on my shirt, my left and right shoe. I would try to swallow a piece of Ivan's tasty bread and Aunt Terez's apricot jam, wrapping my hands around a mug of hot milk. I would hear my grandpa coughing in his bed, and Uncle László hitching a horse to a cart, loading our suitcases and opening the wooden gates.

The last stars would be fading as we were leaving my grandparents' home. As on my arrival, Grandma would be standing at the door, her hair tied and tucked under a grey headscarf. The flickering light from the lamppost on the street didn't allow me to see her face, but I knew she was crying. Uncle László was silent, with the reins in his hands, my mother and sister weeping. As the cart was getting closer to the train station, a tiny ring of red sun appeared on a greying horizon. On the watchtowers of the Yugoslav and Hungarian guards, the lights were still on. Those young soldiers were sleepy and tired, fed up with the military drill and generals, and homesick, I thought. Unlike me, they probably hated their summer on the border in the middle of the Pannonian plains.

ON A FREEZING February day in 1991, I was in Budapest preparing a documentary for Television Sarajevo about the sweeping changes in Hungary. While waiting to interview an expert in a hotel on the banks of the Danube, I observed Soviet generals and high-ranking officers marching through the halls of the luxurious edifice. They were negotiating the final arrangements for the dissolution of the Warsaw Pact. Ironically, this took place in Hungary's capital, where Soviet tanks and troops had in the name of the alliance squashed the 1956 anti-communist revolt.

I witnessed the last gasp of the Cold War. Somehow, I didn't feel any relief or joy. Memories of my happy summers were still stronger than any feeling about the end of the time of danger called the Cold War.

Nikola Gurovic is a freelance journalist living in Brisbane. He worked for Television Sarajevo and Radio Free Europe in Prague and Washington, and his memoir 'Friend for all times' appeared in *Griffith REVIEW 22: MoneySexPower*.

REPORTAGE

Food and prayer

Eating with the locals

Carolijn Visser

Translation: David Colmer

LIKE a Hindu goddess, the scrawny old cook seems to have multiple arms. In no time flat, she's fried the prawns and the orange curry paste and is hurling noodles into her enormous wok. She splashes in a dark-coloured sauce and pumps up the flame. Two eggs fly through the air; shells break on the side of the wok; the contents land in the middle of the noodles. Bean sprouts, spring onions and three squirts of liquid follow. An assistant stands by with a plastic plate covered with a banana leaf. Then a cloud bursts overhead and rain starts drumming on the tin roof. The cook doesn't seem to notice; she's already halfway through the next order.

All the Chinese of Penang know what you mean when you say you're going to eat *char kway teow* at the Sisters'. 'They're so old now,' the driver who drove me to their restaurant said, and sighed. For more than fifty years the two women have been preparing the same dish alongside the same busy roundabout. In the morning, the thin one does her conjuring act; in the afternoon, it's the plump one's turn.

I sit down, wait until they bring me a plate, then raise a mouthful of creamy noodles to my lips; the sauce makes me think of the sea and fishing ports.

That afternoon, I let lunch settle in my hotel room, a tenth-floor eyrie from which I can see the ferry that brought me here last night tracing a line across the bay. Gleaming below are the red-tiled roofs of the old town.

George Town, the capital of Penang Island, is like the China I know only from yellowed postcards. Wide bamboo blinds in front of shop windows protect wares from the scorching sun. The façades are hung with loud vertical banners with advertisements or Chinese characters. Trishaw riders pedal down the streets, and every corner has an old-fashioned restaurant that serves the dishes Penang is famous for. That's why I've come: to eat.

In the course of the afternoon, I descend and cross the street to try some *lorbak* in the Kheng Ping Café. A few groups of people, young and old, are sitting at the round tables on bentwood chairs with tall glasses of fruit juice in front of them. At a small wheeled kitchen, a cook in a white hat deep-fries pieces of pork wrapped in sheets of tofu and serves them to me with a dark plum sauce. I press my arms against the cool marble table-top. Even with the ceiling fans on, the heat is almost unbearable. White light glares in on all sides. My plans for the rest of the day evaporate. 'Why aren't the restaurants air-conditioned?' I'd asked Mr Yap, who had given me my first tips in this culinary paradise. 'Because the owners live off a big, rapid turnover,' he had explained. 'They don't want you hanging around. It's eat, sweat and go.'

I HAD MET Mr Yap that morning. He is the distributor for a Chinese print-er's that I had mistaken for a bookshop and entered in search of a cookbook. Mr Yap hadn't any, but I was in luck all the same. He had a delivery ready for the city's largest bookshop and was happy to give me a ride; there they would be sure to have one. On the way he pointed left and right: 'In this alley you can get a good *sotong kangkung rebus* in the morning – something with squid. Left there and straight ahead, they do a Penang *rojak* that's not bad, fruit with a very special sauce. That woman there has good *chai kuih,* pastries.' Penang is a labyrinth with delicious snacks for orientation.

'Food seems to be the most important thing around here,' I remarked. 'And prayer,' Mr Yap corrected me. That brought us to the touchy subject of the city's ethnic contrasts. Many Chinese had left Penang, Mr Yap said, because they felt discriminated against in Islamic Malaysia.

Nowadays some of them were returning from overseas. 'With a lot of money sometimes,' according to Mr Yap. 'They buy run-down houses in the old centre and do them up.' I had noticed homes like that. They had gilded doors with magnificently restored woodcarvings. When they opened you saw floors of antique Portuguese tiles and inner courtyards with luxuriant palms.

The houses had originally belonged to prosperous Chinese merchants. 'But in the last few decades, many poor families lived together in those mansions,' Mr Yap said. Old Penang had gone into decline and lost its charm for him. Like others, he had moved to Air Itam, a new suburb in the hills where the flats were modern and the air fresh. Mr Yap pointed out a Muslim Mamak restaurant where *nasi kandar* is served. 'That is very good, too.' 'So the different ethnic groups don't necessarily eat apart?' I asked. A broad smile appeared on his face. 'The Malays don't come to our restaurants, because we use pork and dripping. But at theirs we can eat everything!'

WE HAD ARRIVED at our destination: Gurney Mall, with its view out over the sea. The bookshop was on the fifth floor. Mr Yap led me to a shelf and pointed out *Famous Street Food of Penang: A Guide & Cook Book* (Star, 2006), which had recently been published. Before parting, I asked him, 'Would you say that all this delicious food makes the people of Penang happy?' Mr Yap thought for a moment. 'No, it's like anywhere else: people complain that they are always stuck in traffic and everything is getting more expensive.'

I leaf through my new guidebook in my hotel room, with the air-conditioning humming. 'A well-known curry mee stall can be found in the evenings in Chulia Street, just before the junction to Lorong Cheapside. Hawker Tang Twa Chait has been selling curry mee for over thirty years,' I read. 'His recipe was passed to him by his father-in-law, but he has since improved on the taste of the chilli condiment.' Chulia Street is close by; my evening menu has been decided.

The next day I get up early to make the morning market in Kuala Kangsar Road. There are stalls with fresh fruit and vegetables, and others with pans and shoes. It is busy in the Soon Yuen coffee shop. Women with large shopping bags are waiting for *kway teow th'ng*, noodle soup. I go up to a stainless-steel stall and order one with extra duck from a cook in a baseball cap. The owner, I read in my new guidebook, inherited his collapsible kitchen from his mother, who started the business in 1957. I discover that the stock is perfect, the bean sprouts crisp and the fish balls deliciously salty. Then, with my stomach full to bursting, I take a taxi to the botanical gardens. It's time for a long walk.

Behind a moon-shaped gate, a path leads up a steep hill. Within a few steps, I find myself in a tropical rainforest: wide, dark-green leaves unfold; crickets make a deafening racket. A climb of more than an hour brings me to the top of Penang Hill, where I settle down on the terrace of the colonial-style Bellevue Hotel. Dark clouds appear. I drink English tea with milk while veils of rain pour down, blurring the view of the distant bay. When it is dry again and I am descending the hill on the funicular railway, I realise to my great relief that I can already detect the first pangs of hunger.

Carolijn Visser is the author of twenty books, including *Voices and Visions: A Journey through Vietnam, Grijs China, Buigend Bamboe, Miss Concordia, Shanghai Skyline* and *Oom Brian*. She lives in Amsterdam.

David Colmer won the 2009 NSW Premier's Prize for translation..

Between two worlds

A princess finds home in Italy

Hamish McDonald & Desmond O'Grady

IT is chilly in Florence when we get off the fast train up from Rome. The calories from the rushed breakfast, the café latte in the dining car as the sere landscape whizzed by, are fast wearing off as our taxi crosses the river and winds through the narrow lanes of the old San Frediano district of artisan workshops.

In Via Camaldoli, we find a plain two-storey shopfront that bears the nameplate Associazione Culturale Arte e Gastronomia Orientale. Since 1983, the association has provided cooking classes in English and Italian, teaching locals the finer points of Italian and Asian cooking.

The door is opened by a woman of nearly seventy-seven with the fine features of a great beauty, the hair now grey, recognisable from the photographs of a bygone era in a far-off land of sweeping rivers, layers of blue mountains and paddy fields. She is wearing an ankle-length tight woollen skirt in a tartan pattern, with a blue cashmere top.

'It's tartan in your honour,' June Rose Bellamy, also known as Yadana Nat Mai, says with a huge smile. 'I like to dress in Burmese style but you've got to face it, it's cold here!' It's obviously warm, but also quite oriental in effect, a winter version of the figure-hugging sarong and blouse of South-East Asia.

We walk in, past drawings of horses – the passion of her father, an Australian bookmaker – and Asian artefacts; past framed formal portraits in black-and-white and sepia of Asian people in lavish formal dress, and new colour photos of a village aid project in Burma; and past a big table set for lunch, with an enticing orange flan heavy with a luscious-looking sauce waiting on a floral dish in the middle.

We sit in a sofa and armchair under the portraits. But Bellamy is constantly breaking off to disappear into the kitchen. Eventually we follow, into a large space of grey marble chopping surfaces, hanging pots and pans, and a huge stainless-steel cooking range, and settle around the kitchen table.

'Have you tasted really good Tuscan new oil?' Bellamy asks. 'This one comes from the house of a friend of mine, and I would say it's this year's very

---

PREVIOUS PAGE: *Something from Kate*. Photographer: Paul Mallam
from the *Sydney Morning Herald* Shoot the Chef! 2009 photographic competition.

best.' Slices of bread are brought toasted on a metal grid over a gas burner –
'You never get a toaster that gives you a taste of the flame' – and we dip into
a dish of very light green oil.

It tastes light, a little grassy. 'It comes from higher up in the hills, and it's
made by a fanatic. He even removes the stone from the olive: it is made from
just the pulp, and his home growth is of only one type of olive. To be doing it
properly you should be tasting without the bread, just off a spoon, but I won't
put you through that. When I tasted it, I almost lost a friend, because I said
that it wasn't olive-y. It's very good, but I think it's got something to do with
removing the stone. Maybe that was in my head. It's very light.'

IN MANDALAY BURMA of the nineteenth century, the scholarly and
pious Mindon was made king in 1853 with the help of his warrior brother
Kanaung Min, to start last-minute reforms to stave off the encroaching
European powers, as the legendary King Mongkut was doing in neighbour-
ing Siam. But in 1866 Kanaung Min was killed in a bloody takeover attempt
by Mindon's own sons. The Prince of Limbin, a son of Kanaung Min, was
smuggled to safety.

By 1885, as British steamboats packed with soldiers came up the
Irrawaddy to storm the vast Mandalay fort-palace, depose the last king
and exile him to India, the Limbin prince had grown up and become the
figurehead of a rebellious confederacy of Shan princes, who turned from
fighting the Burmese to opposing the British. In 1887, the prince was forced
to surrender when the Shans did a deal with the British, and he too was
exiled.

One of his daughters, Princess Ma Lat – June Rose Bellamy's mother
– was born in Calcutta in 1894, and later attended high school in Allahabad, mixing in high society. She met the German crown prince, the future
Kaiser Wilhelm, who thought her the most striking woman he encountered
on his Asian tour. She was betrothed to the young ruler of Sikkim, but he
was poisoned eight days before their wedding. Further attempts to arrange
a marriage were abandoned. Small and exquisite, with a will of steel, the
princess was a free spirit whom nobody liked to counter.

A NEW OIL is produced. 'This I find a typically good oil: it comes from this area, a little high; it has more body,' Bellamy says. 'The previous one was super-delicate; you tend to get lost. You couldn't possibly cook with it. Even if you used it in a salad or something like that, it would get lost in the food. To me, this is Tuscany. It's not for sale; they only make enough for the house, and they give a bottle to friends; it's very smooth, a nice artichoke taste.' It is indeed very flavoursome, without the sharpness that catches the back of the throat with most olive oils.

AS BRITISH GUNS crushed Burmese independence, Herbert Bellamy, born in 1878 to English migrants, was growing up in Victoria. He ranged across the goldfields as far as Kalgoorlie, making and losing money, augmenting his funds from wins in professional sprint races and organising boxing bouts. He tried running sheep but couldn't stand the smell of them, and concentrated instead on his racehorses. After excursions to London and France, Bellamy became a fixture on an Asian circuit between Bombay, Calcutta and Batavia (Jakarta), as both breeder and bookmaker.

One day he was chatting with the Sultan of Johore in Malaya, a friend with a common interest in dog fighting. 'Y'know, Bellamy, you should go to Burma,' said the sultan. 'It's an interesting country. Maybe you could do something about horses there.' Bellamy went, fell in love with the country and settled at the Rangoon racecourse.

After World War I, the Limbin prince was allowed to return to Burma and his extensive family settled in Rangoon. They were anachronisms in a booming new state of mines, rice exports, railways and the famous steam-boats of the Irrawaddy Flotilla Company. Ma Lat attended the races and went to place a bet, where she met the expat Australian: the attraction was instant.

Maurice Collis, a colonial judge and author, wrote in his 1938 book *Trials in Burma* of receiving a call from Ma Lat in 1928. 'She sat on the sofa, a beautiful woman, in a blue silk skirt and a jacket of white lawn, her complexion corn-coloured, her eyes large and brilliant, and with exquisite hands.' She asked him to wed her and Bellamy.

The marriage took place on 18 October at 4.48 pm, a time fixed as auspicious by an astrologer. 'She was dressed reminiscent of the court of Mandalay in a royal hta-mein of oyster-coloured silk set with silver diamante; her hair, in the loose tail style, was charged with orchids and there were pearls winding at her throat and breast. She came forward slowly, waving a white fan, with a look of dignity and emotion on her face. Mr Bellamy followed in a morning coat. The ceremony of civil marriage is exceedingly bald, for it consists of hardly more than the taking of oaths. When some documents are signed it is all over. On this occasion its abruptness seemed almost rude. I declared them married and took Ma Lat's hand. The occasion seemed to be strange and disturbing. Had the rape of Mandalay ended in this?'

BELLAMY BRINGS ANOTHER dish of dark brown paste, made by her pupils. 'Now this is 110 per cent Florentine, Tuscan,' she says. 'It's made from chicken livers. Traditionally it should be made from the spleen, but you try to get an American tourist to stand there and get blood out of the spleen. So we do it with the liver. This is country food, what you serve before the meal. There is not a single part of the animal which is not used.' She brings out a bottle of Il Principe pinot nero, from a vineyard named after Machiavelli.

JUNE ROSE WAS born in 1932, and given the name Yadana Nat Mai (Goddess of the Nine Jewels) as well as her English one. She began an idyllic childhood between the two cultures of her parents. 'Mummy was still very much within the laws of what was her due, what was her right, as royalty,' she said. 'At home we spoke Burmese – not the Burmese that ordinary Burmese speak, but court Burmese. I grew up speaking that, and English, and Hindi. I lived on one side very Buddhist. We never wore our shoes upstairs. The food for the monks was prepared first thing every morning. When it came the time for the festivals, my father would disappear hunting. From his side, we had Christmas trees, we had Easter bunnies; I rode, and I shot. What Buddhist shoots?'

June Rose was very close to her father, who set up bamboo hurdles for her in the yard, told her yarns about his time in the Australian bush and read her Henry Lawson's poems. 'I can tell you everything about Kalgoorlie, the Southern Cross; I can tell you about the half-bald cockatoo in the pub in whose cup people would pour beer, and when the parrot was sloshed he'd say: "Give me another feather and I'll fly."'

In February 1942, when June Rose was nine, the idyll ended. Japanese bombers raided Rangoon as the prelude to occupation, narrowly missing the Bellamy house, and the family were evacuated to India. They settled in Allahabad, the sleepy North Indian city at the junction of the Ganges and Yamuna rivers, where the religious gathering of the Kumbh Mela, the largest human assembly in the world, occurs every twelve years.

The city was also home to the Nehru family. An aunt, one of the Limbin prince's other daughters, had married an Indian maharajah allied with Jawaharlal Nehru's independence struggle and who was with him in the Allahabad jail. June Rose went along when her aunt visited. 'My uncle was in one cell, and there was water on the floor with the fan blowing on the water to keep cool. Four doors down was Nehru,' she recalls. 'When I saw the film *Gandhi* I had to see it again. The first time I cried all the way through.'

June Rose was sent to a convent school in Kalimpong at eleven, where Herbert Bellamy's influence was her downfall. A nun who clearly looked down on mixed-race children was giving a geography lesson about Australia, and sarcastically asked June Rose if she'd left anything off the map on the blackboard. June Rose put a dot in the centre. 'Baragarawindy,' she told the class, 'is dream country; it is the land of opposites, the rivers flow inland instead of out, the leaves grow upwards instead of down, the snakes have feathers and the crows fly backwards to keep the dust out of their eyes.' She was expelled.

At fourteen, she returned with her parents to a wrecked and bombed Rangoon. Her father went into semi-retirement in the British-built hill station of Maymyo, outside Mandalay, where teak and brick houses in a vaguely Tudor style sit amid lawns and beds of roses and sweet peas. Bellamy ran a few horses, disappeared to his hunting lodge in the forests and collected rare orchids. The shrinking British community looked down on him as a bookie and colonial; they and the Burmese sniffed at mixed-race families; but both communities had to open their doors because of Ma Lat's royal lineage. 'On one hand that put me in a privileged position but at the same time it raised my gall, because I was no better than any other Anglo-Burman,' June Rose explains.

She was growing into a noted beauty. The English travel writer Norman Lewis met the Bellamy family at a party in Maymyo, and in his 1955 book, *Golden Earth*, called Herbert Bellamy 'a man of genial and confidential manner' and Ma Lat a 'still handsome' woman with an expression of inner amusement. June Rose 'allied to the graceful beauty of the Burmese a quite European vivacity'. And practicality: 'When the family were about to leave, in an elderly and ailing British car, June Rose showed much skill in locating a short in the wiring, and much tomboyish energy in winding the starting handle until the engine fired.'

Winning an essay competition, June Rose was given a three-month tour of the United States as a prize, and flourished on what passed for the social circuit in the fragile yet hopeful fifteen years after Burma's independence, in 1946, when democracy faltered amid repeated insurgencies. She was a contender for a role in the war movie *The Purple Plain*, as the young Burmese nurse who gives a suicidal pilot (played by Gregory Peck) an interest in life, but says she pulled out during the shooting in Ceylon. 'It was so Hollywood, it was ridiculous; it was an insult to anything that had to do with Burma,' she said. 'When the film did come to Burma there was a big hue and cry. Things in the pagoda, things a Buddhist would never do.'

In 1954 she married Mario Postiglione, a young Italian doctor working on malaria prevention with the World Health Organization. Seven months after the birth of the first of their two sons, Mario was kidnapped by Burmese communists. 'They were all well-educated – former students,' June Rose said. '"We have nothing against you or him," they said, "and how is your royal mother and your son?" It happened during the visit of [the Soviet leaders] Khruschev and Bulganin to Burma. "We want the world to know the government has no control of the country. We need the money to buy the arms. Too bad it's your husband." After we got him back, the UN told us to get out of Burma.'

June Rose went with Mario to WHO postings in Damascus, Geneva and Manila. From the Philippines, she was able to fly back to Rangoon to be with her father when he died, in 1963.

NOW COMES THE main course. June Rose moves to the oven and pulls out a large covered dish. Inside is a roll of meat, surrounded by small potatoes roasted a dark golden-brown in a coating of oil and pan juices. 'Rabbit – isn't he gorgeous?' she says. 'This is very typical of Tuscany. This is another one of those dishes that make country food interesting, because a little has to go a long way. It's actually a very small rabbit, but he's been pounded flat to make him double his size, and inside you have pancake, the liver minced up and fried with onions. So from three eggs and one rabbit you can feed eight to ten people.' From a bench top, she produces the greens, a circular flan of shredded zucchini. We set to, with topped-up glasses of Il Principe.

IN ITALY, JUNE Rose and Mario divorced, and she was raising their two boys, Michael and Maurice, when she learned that Ma Lat was gravely ill after a stroke. By then, Burma had become isolated after the army chief, General Ne Win, seized power in 1962 and stepped up the ongoing wars against the many domestic insurgents.

'I couldn't get a visa, so I sent Ne Win a telegram – I'd known him for years; I was a very good friend of his wife, Katie; we'd always sort of kept in touch,' she said. 'So I cabled him: "Mummy critical. Rome embassy incapable of giving me a visa. Please help." The embassy called and said I didn't even have to go to them: they'd meet me at the airport. But in the meantime I got a cable saying mother had died. So I didn't go back. I wrote and thanked him. I would have done anything to see my mother, but to go and collect her ashes…

'A year later he came to Europe. And we met. I thanked him for what he'd done. I discovered that Katie had died. And that was how our connection started. He asked what I was doing here. I mentioned my children, my divorce. He said I should go back to Burma. To do what? This thing went backwards and forwards, and on another trip he proposed. I hadn't been back to Burma; all this happened here.' She went to Rangoon in 1978, and they were married. 'It was not one of those lavish things.'

June Rose is distracted, edgy as she talks. She looks for the small bottle of beer she has been drinking instead of wine. She gets up and clears the plates, clattering noisily in the sink.

IN THE LATE 1970s, Ne Win and his military regime were losing international respectability. Earlier, the anti-communist general had been useful to the West, and his isolationist economic policies and authoritarianism not so unusual. Gough Whitlam had even brought him to Canberra on an official visit in 1974.

Now the dance with Deng Xiaoping's China had begun. The rest of South-East Asia was starting to pull ahead. Ne Win's superstition and his fascination with lessons from Burma's, blood-soaked history were becoming obsessive.

But he was still an attractive man, recalls Pamela Gutman, a Sydney scholar of Burma who finished her doctorate on the cultural history of Burma's Arakan state about that time. She presented her visa and an introductory letter at Ne Win's gate, after being studied by a periscope that appeared over the wall and swivelled in her direction. A dinner followed, with Ne Win serving game he'd just shot up in Shan state, along with a bottle of sweet German wine – a taste he'd evidently developed on his annual trips to spas and physicians in Germany and Austria. The conversation was desultory, with the leader consulting a colonel often about what he should say. Bizarrely, it ended with Ne Win suggesting an Australia–Burma cultural agreement. The Australian ambassador was excited. 'What sort of scotch does he drink? I'll send a case in.'

Ne Win thought marriage to June Rose would be advantageous. 'All the locals would say it was a good thing because she had royal blood, and legitimised his regime,' Gutman said. June Rose agrees. 'I think – and people say it, which is why I can say it – I was a sort of lollipop for the people,' she says. 'Whatever average people say about me or my Anglo half, the family name is still very important in Burma, the royalty, the Limbin.'

In her mind, June Rose says, was the idea of doing some good for the country. But she admits failure, though she won't talk about any of the conversations or issues she had with Ne Win. She suggests that he was barely in control of his regime by that point.

'Had I been able to do anything, had it served a purpose, had something been able to be done – but I realise I saw too many things,' she says. 'He was being taken for a ride by his people. We're not talking about manipulation, but being put in a position where you don't know everything. You think you know everything, but a dictatorship in a country like Burma, as long as it is, with all the different tribes – each command is a watertight compartment.'

Ne Win never felt he was out of the loop. 'Oh, they weren't that stupid. Neither was the media, which needed a scapegoat. He was always the *éminence gris*. He was always the one who was manipulating everybody. It was too convenient,' she said.

'But all you have to think is that when he died [in 2002, after being sidelined by his army peers in the wake of the 1988 uprising], he died alone, abandoned, ignored. And something that would be impossible for a normal Buddhist or a Burmese to conceive: he was buried the same day without a friendly soul by the graveside. After all that time? What were they hiding? Why did they have to get him out of the way as fast as possible? To cover up things. They did what the British did – do you know my grandfather's grave didn't have a name on it? He was to be forgotten.'

The marriage ended after just five months. June Rose won't go into the cause of the final rupture. Gutman says the rumour is that Ne Win was entertaining one of his wartime Japanese mentors, who was by then working for a trading company, when June Rose mentioned the worsening state of Burma's economy. June Rose laughs at another popular rumour, that Ne Win suspected she was a western spy.

She does confirm that it ended when Ne Win threw an ashtray at her. 'It was one of those things that happened. Rage. Anger. I know why, but on the other hand I don't. It's something that is too questionable. The fact remains that yes, there was a physical attack, but even that is not simple,' she says. 'Okay, so he did fling the bloody ashtray. I can't deny it, because there were servants and obviously it is the servants who are talking today…There was an ashtray, but it didn't hit me between the eyes and I'm still alive. But it's not the ashtray. It is the last drop in the glass, the last straw.'

June Rose left the next day, seen off by Ne Win's daughter, Sanda (now jailed for corruption by Ne Win's successors), and a guard of honour as she flew out in what was then Burma's only passenger plane certified for international routes. Word of the marriage breakdown had not spread. Ne Win had gone up country. She felt lucky to get out.

'I LEFT BURMA with a definite feeling of failure,' June Rose says now. 'Because I had failed my people. Because they did put their trust in me when I arrived. And this was one of the things that was not liked. But I would rather I left as a failure than to be connected with the ruling people. Those who had trusted in me, those who believed in me can say she left, but she left rather than not be able to do anything. In Italian, they say *un peccato di orgoglio*; in English, a sin of pride. Because I thought I could do something which others had not done. And that's a very bad sin.'

June Rose has written an engaging essay about her family and upbringing, but says she won't write about her time with the dictator. 'To say something, to write a book, one would have to have a very good knowledge of Burmese history, culture, and a super-excellent knowledge of the construction of a military dictatorship,' she said. 'You can't put all that in a book. You'd bore people to death.'

Gutman says that there was talk among diplomats of the time that June Rose had indeed written, or was about to write, a book titled 'One Hundred Days with Ne Win' but was persuaded not to publish by a sizable settlement with Rangoon. In Florence there is comfort, but no sign of wealth. Her shopfront school of Burmese and Italian cooking, with her apartment upstairs, provides a modest living and is gaining a reputation on the self-education tourism circuit. Her charitable work, through Rangoon-based doctors, is putting young Burmese students through medical school and helping a village hit by last year's Cyclone Nargis.

Her younger son, Maurice, died in a motor accident some years ago; but his son, Alex, twenty-five, is now close to her, as is her older son, Michael. Two years ago, she and Alex went to Burma and found the grave of her grandfather, the Limbin prince, in a monastery. At 5 am, when the earth's energy is said to be at its peak, they paid homage by serving the morning meal to the ten abbots and 120 monks. Alex, who had never been to Burma and until then seemed completely Italian, said he felt at home. 'I didn't feel I was visiting,' he told her. 'Everything we do at your home in Florence, people are doing all around us here.'

June Rose says she still tries to lead a Buddhist life. 'But the other day we went to a fun fair, and I was still able to topple off, without my glasses, a barrel at the shooting range.'

We eat the orange flan. It is soft and creamy, with a heavy flavour like curaçao liqueur.

Suddenly we realise it is mid-afternoon. We leave the half-world of Burma, pioneering Australia and Cold War diplomacy that the lunch has conjured up, and step into the subdued light of wintry Tuscany. June Rose farewells us, beaming as she did on our arrival.

Hamish McDonald is the Asia-Pacific Editor of the *Sydney Morning Herald* and a former correspondent in Jakarta, Tokyo, New Delhi and Beijing. As the political editor of the *Far Eastern Economic Review*, he wrote an article on Burma that was read into the record of the US Congress.

Desmond O'Grady has published more than a dozen books, and claims to be the only Australian author to have played cricket at Lord's and tennis with Rod Laver, and made a film with Claudia Cardinale.

THE WEDDING SPEECH

JEREMY CHAMBERS

IT was a long time ago and we were unemployed. Me and Mike were sitting in his carport drinking. That was what we did, back then, most nights, when we were unemployed. Problem was we had Ronald Stott with us, and he was getting on our nerves. But Ronald Stott always got on our nerves.

We had never liked Ronald Stott, not since we were kids and he used to show off his Gray-Nicolls cricket bat but wouldn't give anyone else a go. The same went for when he got a remote-control car and a ten-speed bike. Despite the fact that Ronald Stott was spoilt, or probably because of it, none of us had noticed that his parents weren't all that well off. They weren't poor; they just weren't well off.

I'd been in school with him, primary and secondary, and during all those years Ronald never had a single friend. I occasionally recalled the lonely figure in the school ground at lunchtime, standing in an attitude of dogged pride that seemed to say his solitude was a matter of choice. Years later, I'd occasionally take pity on Ronald and try to be a friend to him. But it never lasted long. In fact, after spending time alone with Ronald Stott, I disliked him all the more.

After school, Ronald Stott had gone on the dole and taken up bodybuilding. He liked to show off his swollen body in tight T-shirts and singlets and would use any excuse to take his top off, especially if there were girls around. He started hanging around with a group from the local football club. They spent most of their time outside the pizza place down the shops, watching girls. Some nights they went down to the toilet block behind the library, where they bashed men who met there for sex. After a while they got sick of Ronald, same as everyone did, and that's when he latched on to me and Mike.

I sometimes wondered why Ronald Stott spent so much time with the two of us. Maybe it was all those years he spent alone at school, because he never seemed to take much interest in anyone apart from himself. But he did like to talk, and he talked a lot, and that was what he talked about — about himself. He'd tell us detailed stories about his conquests when out with his footballer mates: sometimes at nightclubs or parties but mostly in houses and garages and gardens and lanes, with girls they joked about afterwards. The local bikes, they called them. He would recount his daily gym sessions: the order of exercises and lifts and the weight lifted; the number of repetitions; how knackered he felt at the end of it and the quantity of food he ate when he got home. He repeated the remarks he made to the old Vietnamese couple who shuffled the streets delivering junk mail. He used to stand in his front yard every afternoon, waiting for them to come past. Ronald Stott didn't really have a great deal to say, but he took a long time saying it. We still didn't like him, he was still spoilt, and his family still weren't well off.

In fact, it was around that time things were starting to fall apart for Ronald's family. His father, a small, angry man with thick-rimmed glasses and a passion for racing cars, had recently lost his accounting job after years of liquid lunches finally caught up with him. He now worked as an itinerant bookkeeper for independent car dealers, and Ronald's mother, as Ronald had no qualms in telling us, was threatening to leave him if he didn't stop drinking and get a decent job. However, none of this seemed to affect Ronald, at least not so far as we could see.

So anyway, the three of us were drinking in the carport when Jenny and her friends pulled up opposite Mike's place. Jenny was Tank's fiancée. Tank was a good friend of ours.

It was late and they had been out somewhere and were dressed up. Jenny and one of her friends got out of the car. I forget her name: the friend, I mean. They stood by the car, caught up in conversation. The two of them were in hysterics over something.

Shelley, who we'd known for a while, was still inside. She was leaning against the door, her face flattened against the window and pale in the streetlight. She was wearing pink lipstick and pink lipstick was smeared across the glass.

Aw, Christ, said Mike. There's Shelley.

What's wrong with Shelley? I asked.

It's nothing, said Mike. She's just been a real pain in the arse lately. Don't worry about it.

You don't look at the fireplace when you're stoking the fire, said Ronald, laughing.

Mike moved his chair over so he was no longer facing the street. It scraped against the concrete and he put his feet up and sank down and sighed.

Shelley's all right, I said. What's wrong with Shelley?

Jenny's friend was calling us over from the street.

Ignore them, said Mike, but me and Ronald went over. Ronald kept nudging me as we walked. I pushed him away.

The friend had eyes shiny with booze.

Do you know what Jenny did? she asked us, pointing at Jenny.

Jenny screeched and tried to grab her friend, but the friend twisted away.

She took her engagement ring off, the friend said.

I did not, Jenny said. Don't tell them that. She started grappling with her friend and making a hell of a racket. They were very drunk. I didn't really want to be around them and I shouldn't have been around them anyway. Not when they were like this, acting like this and drunk and whatever.

The thing was, Tank had asked me to be best man at his wedding and I had to make a speech. I knew that in a wedding speech you have to praise the couple and talk about their love and how they are made for each other, how you knew from the start they would be together forever and that sort of thing. Not that I'm saying it's a lie, but it's not the truth either. Truth is a complex thing and it has no place in a wedding speech.

Yes, you did, the friend said. She was smiling. She had been smiling the whole time.

She put it in her handbag, the friend said to me, looking at me, still fighting off Jenny. She took it off before we even went in.

Yes, you did, she said to Jenny. I saw you. We all saw you.

They kept pushing and grabbing each other. Jenny was trying to put her hand over her friend's mouth. They were wearing tight-fitting dresses, stockings and high heels. The friend's dress was small, red and strapless. Her skin was very white. Wherever they had been it must have been somewhere nice.

Ask Shelley and Michelle, the friend said to me, ducking out of a hold. We all saw it.

Ronald was walking around the car. He put his face against the windows to look inside. The windows were steamed up and the metal was already damp and dripping.

It was a cold night and winter. I had said to Tank that it seemed like an unusual time to have a wedding. Tank told me that Jenny had her heart set on a picturesque church in Brighton on an elm-lined street and not far from the beach, where the wedding photos were to be taken. The problem was that she wasn't the only one who had her heart set on that church and it was heavily booked. Jenny wanted the perfect wedding.

She wants the fairytale wedding, Tank had said grimly.

Ronald leaned over and wiped the moisture off one of the windows with the sleeve of his jacket. He waved to Shelley, who was sprawled across the back seat. She peered through the misted window and tapped it and said something I didn't hear.

Mike had come over by then. He had a stubby in his hand and was looking down the street, away from the car and the girls and their shrill melee. Mike was making it clear that he didn't want to be there.

So what's going on? he asked me.

I watched Jenny and her friend. Their bickering was beginning to irritate me. Mike drank and looked about as though he were standing alone and lost in his own thoughts. Eventually Jenny got hold of her friend and dragged her away from the car. The friend was calling out in a singsong voice.

We saw you. Everyone saw you.

Mike looked disgusted.

The friend was bent over, her head locked in Jenny's fleshy arm. She had to lift her face to look at us.

Are you going to tell Tank? she called out. Are you going to tell him about the ring?

Well, we're going to have to tell him now, I said. Before he reads it in the papers.

The friend looked at me, slightly surprised. She didn't get it.

Jenny suddenly lurched at me and grabbed me around the shoulders. She hugged me and rocked against me, pleading in a small and plaintive voice. I felt very uncomfortable. I was feeling uncomfortable about the whole thing, especially given I had to make that speech.

All right, all right, I said.

Shelley had wound down the car window and was calling for Mike. Her voice was slurred and she looked half asleep. Ronald was standing next to her with his thumbs tucked into his belt and a grin on his face.

Mike. Come over here, Mike, Shelley said. She held a hand out the window and then let it flop against the car. Her head dropped against the door and she kept mumbling Mike's name.

Jenny had her hand cupped over my ear and was leaning all her weight against me. She was a large girl who smelt of alcohol and perfume and sweat.

Do you want to know a secret, she whispered. Do you know what Shelley said?

No, don't tell them, said the friend, giggling. She was quite composed now and stood with her arms folded, shivering slightly. She looked good in that red dress.

Jenny was still leaning against me but now she was talking to Mike.

Shelley wants to kiss you, she said.

Mike sniffed and looked away.

She said more than that, said the friend.

We had only met Tank because Jenny and Mike lived opposite each other. The first time we ever saw Tank was also the first time he took Jenny out. As we had a view of her front door from the carport, we felt as though we'd watched the whole courtship: from polite and slightly nervous goodbyes to pecks on the cheek to passionate kisses and long embraces. We watched and we speculated and somehow we felt part of the whole thing.

At some point Tank had started nodding to us on these nights, after he had made his farewells and was getting into his car. One night he joined us for a beer.

Tank was originally from the country somewhere and we were the first friends he made after moving to Melbourne. Tank had left school when he was young and worked a number of years laying

sleepers for some railway out in the desert. He had come to the city to
train as a policeman, which he'd done, but he didn't last long in the job.
We didn't see him a lot nowadays. Even so, he spent more time with us
than anyone else, certainly more time than he spent with Jenny. I don't
recall where he was that particular night.

No, she didn't say that, Jenny was saying to her friend.

She did, said the friend, nodding with her eyebrows raised.

Yeah, but not exactly, said Jenny. Not in those words.

Somewhere above us a possum let out a long rattling growl. I
went over to Mike, who was still paying no attention to the girls and
looking nowhere in particular. Ronald Stott was leaning against the
car with his arms crossed and his head through the open window.
I saw Michelle sit up, brush back her hair with both hands and say
something to him.

What's going on with you and Shelley? I asked Mike.

Mike took a swig from his stubby.

Nothing, he said. Nothing's going on. It's just, I don't know.
Tank's been on my back about her. You know, it's all been Shelley
likes you, why don't you ask Shelley out sort of thing.

Mike belched.

Tank thinks just because he's getting hitched, everyone should
be pairing up.

I looked over at the car where Michelle was sitting and listening
to Ronald Stott, nodding. Michelle was a strange sort of girl, willowy
and soft-spoken. She made slow, tired gestures when she talked and
never seemed particularly interested in anything much.

We listened to Jenny and her friend arguing about what was said
and how it was said and then they started talking about something else.
The sky was cloudy and bright, a dirty shifting lavender colour.

But come on, said Mike. Shelley? Are you serious?

I sort of like Shelley, I said.

Well, go for it then. You'd be doing me a favour.

Mike looked at his watch.

You really want to hang around with this lot? he asked me. Why
don't we just piss off?

I looked around at the girls in time to see Jenny run over to the nature strip and throw up against the trunk of a plum tree. Mike's street was lined with plum trees on both sides and in early spring every one of them was a mass of pink blossoms. It was quite a sight.

The friend ran after her, holding Jenny's forehead and gathering up her hair as she kept vomiting. I think the friend was a nurse. Jenny was also a nurse, and it was because of her unusual hours and long shifts that she and Tank never had the chance to spend much time together. That's what Tank told me, anyway.

Jesus Christ, Mike muttered.

A car door slammed and Shelley was out of the car, unsteady on her feet.

Mike, she said. Come over here and talk to me, Mike.

Shelley always dressed too young for her age and talked and acted like a teenager. She had been going to a solarium for the wedding, and her face and arms and bare legs were all evenly tanned. She was short and broad-shouldered and sometimes mistaken for a boy.

Jenny had finished throwing up and the friend was talking to her and stroking her hair.

But I do love him, Jenny was saying, still down on her knees. I do, I really do. Dripping trails of saliva hung from her mouth down to the grass and she was close to crying. The friend kept stroking her hair and making shushing noises.

Shelley was sidestepping in our general direction. She was wearing a tiny sequined skirt, plastic jewellery and a sort of fancy singlet.

I've got something to tell you, she was saying. She was saying it to Mike.

Jenny had lain down on the grass, which was wet with dew. It glistened under the streetlight. No, it's all right, she was saying to her friend, who was trying to lift her up. Just a nice little snooze.

I heard the car door slam again and watched Ronald Stott walking away with Michelle. Ronald was holding her tight against him and she was stumbling a bit. Michelle was wearing a long pale dress, loose and shimmering as she walked. A strap kept falling off one of her shoulders and she kept putting it back. Ronald was talking.

Was that the last beer? I asked Mike.

Last of ours, said Mike.

I went back to the carport and opened the outside fridge. It was stacked with longnecks belonging to Mike's old man. If I took one, I would cop it tomorrow, but I thought I deserved another beer right now, because next weekend I was going to have to give a speech full of high praise and beautiful sentiments about Jenny, who at that moment was passed out in her own vomit.

From the carport I watched Shelley trying to embrace Mike with her lips pursed. Mike backed away and shoved her. He was still holding onto his stubby. Shelley fell, sitting hard on the ground. She looked surprised for a moment and then she burst into tears, hiding her face between her arms. Mike swore and lobbed the stubby down the street. It went high and smashed loudly. Some birds started to sing out of the darkness and stopped abruptly. Mike hadn't thrown the bottle far, and from where I was standing I could see the long glittering trail of brown glass and the slow movement of froth edging towards the gutter. Mike walked across the street and went into his house, slamming the door.

The friend went over to Shelley. I put the longneck back in the fridge and closed the door. The two of us got Shelley up and into the car. We lay her across the back seat. The friend sat on the seat and Shelley sobbed into her lap. She left the car door open.

So what have you guys been up to tonight? she asked me.

You know, I said, shrugging my shoulders. Same old, same old. Yourself?

It was good, she said. It all went off pretty well. Jenny had a bit too much to drink, but it was her night after all.

She looked over in the direction of the nature strip and the plum tree and Jenny sprawled out and sleeping.

Last night as a free woman and all that.

Oh, was this the hens' night? I said. Was that tonight?

Yeah, said the friend. Didn't Tank tell you?

No, I said. I haven't spoken to Tank for ages. Probably been busy with all the wedding stuff, his rellies coming down and everything.

Shelley started saying something in a small voice, whimpering.

The friend stroked her hair and made a shushing noise like she'd been doing with Jenny. She said something about Mike and Mike being a bastard and not worth it and Shelley burst into tears all over again. The friend watched her for a bit.

Well, I hope you boys are going to behave yourself anyway, she said. And make sure that you get Tank to the church in a decent state. I'm serious about that. You don't want to spoil Jenny's day.

What? I said. You mean the bachelor party? Yeah, well. You know Tank. He's the responsible one.

So what's the plan? she asked. Strippers and that sort of thing, I suppose.

Nope, I said. Not at all. Tank said if I hire a stripper he'll walk out. So definitely no strippers.

The friend looked up at me.

So how come you had to organise it?

Well, I haven't really yet, I said. But yeah, I will. I mean, it's supposed to be my job, isn't it. Best man and everything.

The friend closed her eyes and nodded.

Oh, of course, she said. That's right. I forgot you were the best man. For some reason I keep thinking it was going to be one of Tank's old mates. From back home.

No, Tank asked me, I said. Besides, none of them were able to make it.

The friend looked out the window at Jenny, who was lying on her back, her stomach rising and falling. Her dress was hitched up, showing heavy thighs and white underwear under her pantyhose.

I better get her inside, said the friend. She was still stroking Shelley's hair. Shelley had gone quiet. I thought maybe she'd gone to sleep as well. I was looking at the friend. Her skin was very white and you could see faint blue veins underneath. She still looked good in that dress.

Oh well, she said, still watching Jenny. Doesn't hurt to let your hair down once in a while.

Well, you seem to be holding up all right, I said.

What? Me? she said. I'm driving. I'm the designated driver.

Really? I said. I thought you were completely out of it earlier.

No, I wasn't, she said.

Well, you were acting like it, I said. I mean, you seemed to be.

I was just happy, she said. She tossed her head. Her hair was blonde and cut in a bob. It swung and settled.

I was high on life.

Sure you were, I said.

The friend looked at me and scowled.

What's that supposed to mean? she said.

She started attending to Shelley again, licking her fingers and wiping away the mascara that had run down Shelley's face. She looked back up at me, not happy.

I mean, I think I should know what I've been doing better than you, thank you very much, she said. You should spend more time worrying about yourself. You and your mates.

The friend was going to be one of the bridesmaids and Tank had told me that I would be paired up with her in the wedding party; that we would be walking together arm-in-arm during the wedding procession and then I would have to dance with her at the reception. I don't think I knew her name back then either.

THE FOLLOWING AFTERNOON me and Mike were smoking cigarettes in his carport when Tank's Ford Escort went right past us. We could see Tank through the window. He didn't even glance in our direction. He was staring straight ahead and driving fast.

Where's he off to? I said to Mike.

A few minutes later the Escort came back and pulled up outside Mike's drive, breaking hard. Tank got out and I was about to greet him when I noticed he was red in the face and looking angry as hell.

Tank had told me why he left the police force. He said it was because he couldn't wear the uniform. He said he couldn't stand the way people looked at you when you were in uniform and that most people, ordinary people, seemed frightened when they saw you on the street. He said other people would look at you with genuine hatred.

Tank was a thoughtful and decent guy and he said he hadn't been able to hack it anymore, because the minute he put on that uniform he was a cop and nothing else, not himself anymore. He said that when he was in uniform he didn't feel like a person at all.

Eventually Tank had ended up working with retarded kids, and he was always saying how fulfilling it was and that he'd made the right decision when he quit the police force.

Tank came up Mike's driveway glaring at us. He stood with his hands in tight fists and his muscles tensed. Tank had the thickest arms and legs I'd ever seen. His forearms were as big as my calves.

Where's fucking Ronald Stott?

What's up with you? I asked.

He was here with you last night, wasn't he, Tank said. When the girls got back from the hens' night.

Yeah, well it's a whole new day now, isn't it, said Mike. He didn't seem to notice that Tank was all worked up. Mike was lying back on a banana lounge with his feet up and the pale winter sun on him. He was wearing sunglasses and he yawned and stretched.

A glorious new day, he said.

Tank shook his head and started to say something but turned and marched off. When he got to his car he looked back at us, shaking his head again. He opened the car door and then turned back and pointed at us.

And I've had it with you two, he yelled. I've had it up to here. I mean that. You're fucking wastes of space, the fucking pair of you.

Yeah, well fuck you too, Mike called out lazily.

I rang Tank up that night.

You just tell Ronald Stott that when I find him I'm going to knock seven shades of shit out of him, Tank said. And that's not a threat – it's a promise. You tell him that, all right. Just tell him that.

Then he hung up on me.

A similar thing happened a few days later. This time it was Shelley's car that pulled up across the street. Shelley and another woman got out, a woman we didn't know, and they started walking towards Jenny's house. I saw Shelley saying something to the woman, glancing in our direction. The woman turned and looked at us and then started walking across the road with Shelley trying to pull her back.

The woman stood on the pavement with her arms crossed and stared at us. She looked older than Jenny's other friends and had a hard face. She was wearing large sunglasses, so I couldn't see her eyes. After a while, she began talking in a slow, measured way. She reminded me of a schoolteacher. Maybe she was a schoolteacher.

You know Michelle should go to the police about your mate, the woman said. He should be locked up.

What mate? said Mike, opening a bottle.

The woman turned back and asked Shelley something. Shelley was standing on the nature strip, looking at the ground. She seemed embarrassed, or maybe she just didn't want to be there.

This Ronald character, the woman said. Your pig of a mate Ronald. I mean, it is beyond belief. I cannot even begin —

He's not our mate, Mike said, flicking the bottle top over his neighbour's fence.

The woman talked to Shelley again.

He is your mate, she said to us. You were with him the other night. You were seen. And from what I've been hearing, that wasn't the only thing that went on. And I'm talking about the two of you. So you just listen to me for a moment.

It's got nothing to do with us, said Mike. It's her word against his.

But she was passed out! In some lane!

Not when we saw her, said Mike.

The woman held up her hands, swore, started talking to Shelley again and then back at us.

Well, I think you should all be locked up, she said. The lot of you.

She made a sweeping gesture with her arm and then started to storm off. I got up.

Oh, come on, said Mike. You've got to be kidding me.

I caught up with Shelley and the woman outside Jenny's house.

Shelley wouldn't look at me. The woman started talking again. Now she was talking very fast and she was angry and she talked all over the place. She said a lot of things but I couldn't get a word in.

When the woman finally stopped talking she held up one hand with the palm facing me and walked away. Shelley followed her and I went back to the carport.

What a bitch, I said to Mike.

I told you, said Mike. But you've always got to be the good guy, don't you?

IT WAS SOMETIME in the middle of all this, sometime during that week before the wedding, somewhere in the middle of it all I remember a strange feeling coming over me. It weighed on me for a long time, for what seemed like a very long time. And it wasn't anything to do with what Ronald did or might have done. It was something else, but I couldn't quite work out what. Something about that night, for sure, but it was more than that. For whatever reason, I wondered whether maybe Tank had been right to take things out on us, even if it was Ronald he was actually angry at. I remember thinking that maybe it had been a long time coming. I don't know, but something was wrong and I wondered whether maybe there was something wrong with me, something everyone else saw but I didn't and couldn't and can't even now.

Anyway, it was a really nice wedding. I was best man and I gave a speech about what exceptional and admirable individuals Tank and Jenny were, and how they had a love that I knew would endure for the rest of their lives and long after. I said it was an eternal love, and that destiny had brought them together and they were truly kindred spirits and it was rare to see two people with such a close connection and I said other things too and at the time I meant it all. I was probably just caught up in the whole wedding thing because it was a really nice wedding and I think everyone felt very close. I danced with Jenny's friend and we actually got along really well and I danced with her more than once. Mike spent some time talking to Shelley and I remember seeing Shelley laugh at something he said. And I had to admire Michelle, because looking at her you wouldn't have known what she'd been through and she didn't seem to hold a grudge against either of us.

Of course we didn't see much of Tank after the wedding, given that he was a now a married man, but also because he and Jenny bought a house in the outer suburbs and started having kids almost immediately. I haven't seen Tank in many years now, but I often think that I must catch up with him some time, or at least give him a ring to see how he's doing and where life's taken him and that sort of thing.

As for Ronald Stott, our dislike turned to outright hostility and we used to trade insults on the street, especially when he walked past Mike's carport. Mike and I told Ronald Stott exactly what we thought of him and had always thought of him and I remember Ronald looking surprised and hurt. I also remember wondering whether the whole thing had finally touched a nerve, broken through that self-absorbed shell he'd lived in all his life, and that somewhere in that boastful, spoilt brain of his, Ronald suspected that maybe he had actually done something wrong. I doubt he ever would have admitted it to himself, but I do think something had got in there, finally. At any rate, Ronald got a job in computing soon after and moved out of home and we never saw him again.

Me and Mike both eventually found work and moved out of our parents' houses and to different suburbs. We still have the occasional get-together over a few drinks and invariably spend most of the time reminiscing about the good old days. Whenever the subject of Tank's wedding comes up, which it often does, we both agree that of all the weddings either of us has been to, Tank's wedding was by far the nicest.

Jeremy Chambers lives in Melbourne. His debut novel will be published this year by Text.

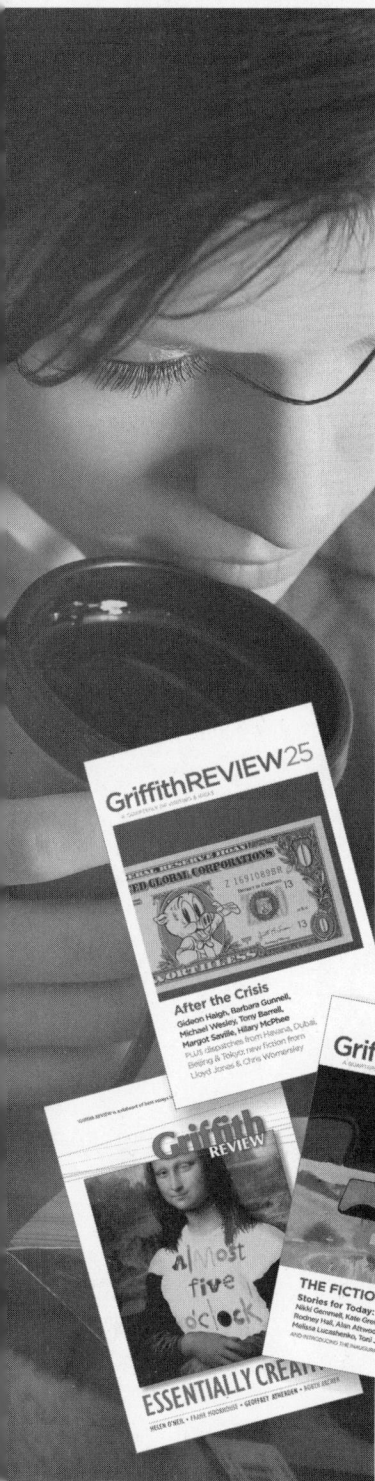

Save 20% with a 1 or 2 Year Subscription plus receive a FREE copy of a past edition of your choice*

☐ I would like to subscribe   ☐ I wish to give a subscription to: *(please tick ✓ one)*

Name: _____

Address: _____

_____ Postcode: _____

Email:_____ Telephone: _____

Please choose your subscription package *(please tick ✓ one below)*

☐ 1 year within Australia: $80.00 (inc gst)   ☐ 2 years within Australia: $150.00 (inc gst)

☐ 1 year outside Australia: $130.00 AUD   ☐ 2 years outside Australia: $250.00 AUD

I wish the subscription to begin with *(please tick ✓ one below)*

☐ CURRENT EDITION†   ☐ NEXT EDITION

For my FREE copy, please send it to ☐ me   ☐ my gift recipient *(please tick ✓ one)*

EDITION TITLE* _____

Select from past editions at www.griffithreview.com ***** *While past edition copies remain in stock.*

PAYMENT DETAILS

Purchaser's Address *(if not the subscription recipient)*:

_____ Postcode: _____

Email:_____ Telephone: _____

☐ I have enclosed a cheque/money order for $_____ made payable
to **Griffith REVIEW** (Payable in Australian Dollars only)

☐ **Card Type (*please circle one*):** Bankcard / Mastercard / Visa / Amex

Card Number: ☐☐☐☐ ☐☐☐☐ ☐☐☐☐ ☐☐☐☐

Expiry Date: __ __ / __ __

Cardholder name: _____

Cardholder Signature:_____

MAIL TO:
Business Manager - Griffith REVIEW
REPLY PAID 61015
NATHAN QLD 4111 Australia

FAX TO:
Business Manager - Griffith REVIEW
07 3735 3272 *(within Australia)*
+61 7 3735 3272 *(International)*

● The details given above will only be used for the subscription collection and distribution of Griffith REVIEW and will not be passed to a third party for other uses. For further information consult Griffith University's Privacy Plan at www.griffith.edu.au/ua/aa/vc/pp ● † Current Edition only available for subscriptions received up until 2 weeks before Next Edition release date. See www.griffithreview.com for release dates.

ED27 1219251109